W0231952

Springer-Verlag Berlin Heidelberg GmbH

Keith J. Gooch   Christopher J. Tennant (Eds.)

Mechanical Forces: Their Effects on Cells and Tissues

 Springer

Keith J. Gooch, Ph.D.
MIT
Department of Chemical Engineering
77 Massachusetts Avenue, E-25-342
Cambridge, MA 02139
U.S.A.

Christopher J. Tennant
6442 Amherst Avenue
Columbia, MD 21046
U.S.A.

ISBN 978-3-662-03422-4
Biotechnology Intelligence Unit

Library of Congress Cataloging-in-Publication Data

Gooch, Keith J.
Mechanical forces: their effects on cells and tissues / Keith J. Gooch, Chris J. Tennant.
p.cm. – (Biotechnology intelligence unit)
Includes bibliographical references and index.
ISBN 978-3-662-03422-4   ISBN 978-3-662-03420-0 (eBook)
DOI 10.1007/978-3-662-03420-0

1. Animal cell biotechnology. 2. Cell culture. I. Tennant, Chris J. II. Title. III. Series.
TP248.27.A53G66 1997 599'.0876041–DC20 96-26773 CIP

© Springer-Verlag Berlin Heidelberg 1997
Originally published by Springer-Verlag Berlin Heidelberg New York in 1997
Softcover reprint of the hardcover 1st edition 1997

Typesetting: Landes Bioscience Georgetown, TX, U.S.A.

SPIN 10628850   31/3111 – 5 4 3 2 1 0 – Printed on acid-free paper

Acknowledgments

This book benefited from the input and careful reading by several individuals other than the authors. Many of the important concepts in chapter 2 were suggested by John Frangos. Mechteld Hillsley provided additional insights which were incorporated into chapter 3. I would also like to recognize the constant support and encouragement of my parents throughout all of my undertakings. Finally, I would like to thank Ling Eng Ang for her friendship and support.

Keith J. Gooch

PREFACE

All cells are constantly exposed to physical forces. These forces modify the biological behavior of the cells, affecting phenotype, gene expression, metabolism and secretion of paracrine and autocrine factors. Though some cellular responses appear to be ubiquitous, especially on the level of second messengers, the application of physical forces often results in well-coordinated responses, such as upregulation of one response with a concurrent downregulation of antagonizing responses. These mechanical-induced cellular changes are manifest as important factors influencing the physiological and pathological condition of the organism.

Cell culture is an important model of in vivo systems and tool for production of commercial products. It is for individuals that use cell culture in these capacities as well as those interested in the physiological and pathological effects of physical forces in vivo that this book is primarily designed. In light of the effects mentioned above, physical forces must be considered in cell culture. When trying to reproduce living systems in the laboratory, important factors such as pH, temperature and nutrient supply are usually carefully considered and regulated. But inadequate understanding of the significance of physical forces and difficulty in generating appropriate physical forces in vitro means that they often are not correctly reproduced. The lack of appropriate physical forces may seriously compromise the validity of cell culture models. Alternatively, when the appropriate forces are introduced, cell culture can be an important tool for elucidating the effects of physical forces in physiological processes such as growth and development and pathological conditions such as atherosclerosis, osteoporosis and osteoarthritis. In cell-based production systems used for biotechnology applications it often is desirable to maximize production of specific compounds. If the mechanical environment of cells was better understood and regulated, production could likely be increased. In tissue engineering where cells or the tissue they form are the desired product, mechanical forces may decisively affect the structure, function and quality of the tissue-engineered product.

This book focuses primarily on the biological responses of mammalian cells and tissues to physical forces and how these responses are related to physiology, pathology and the production of commercial products. The introduction explores the origin of mechanical forces in vivo and methods of reproducing appropriate mechanical environments in vitro. Subsequent chapters are devoted to the effects of mechanical forces on the cellular and tissue level in the endothelium, bone, cartilage and muscle. These tissues were chosen because of the sizable body of literature on their responses to physical forces and their involvement in clinically significant conditions. Within each of these chapters, specific experimental results are presented, primarily to illustrate major points. Many references are provided, among them extensive literature reviews, for readers who wish to learn more about specific aspects of the cellular response. In addition, current hypotheses of potential mechanisms mediating the transduction of mechanical stimuli to initial biochemical responses are presented and critiqued. The book considers the implications of cellular responses to mechanical forces in production systems used in biotechnology applications and tissue engineering. The final section of the book overviews the major concepts presented in the text.

CONTENTS

CHAPTER 1

Introduction

Animal Cell Culture

Animal cell culture began in 1912 when Alexis Carrel grew bits of chick heart in vitro by placing them into a drop of horse plasma. When the plasma clotted, it formed a solid surface into which the heart cells explanted. Left unattended, these cells died within several days. By regularly feeding the growing cells with aqueous extracts of whole chick embryos and periodically subdividing them, Carrel was able to maintain the cells for extended periods. From these initial experiments, cell culture has expanded into an important component of biological research and commercial production.

Animal cell culture is an important tool for the study of complex biological systems. In vivo it is often impossible to target a treatment to one specific cell type or to adequately control the environment. For example, it is impossible to selectively depolarize endothelial cells in vivo to determine the role of transmembrane potential in the transduction of fluid mechanical forces generated by blood flow to a biochemical response. Depolarizing all of the cells within the vascular system by injecting potassium chloride into the animal's bloodstream clearly is not an option, as it will lead to cardiac arrest. Transferring cells from their in vivo environment to a relatively simple in vitro environment facilitates the isolation and control of the variables to be tested. This increase in control, however, is often at the expense of representativeness. It is impossible in vitro to replicate all of the conditions found in vivo, and it is possible, even likely, that many important conditions will differ between the two systems.

Even when it is technically feasible to conduct in vivo studies, it is often more economical to conduct experiments using animal cell culture. In such situations it is possible to evaluate, relatively inexpensively, a number of preliminary hypotheses in vitro and later test the most promising hypothesis in vivo.

Animal cell culture is utilized in the production of many compounds used in health care and related industries. Since the 1950s, animal cell culture has

Mechanical Forces: Their Effects on Cells and Tissues, by Keith J. Gooch and Christopher J. Tennant. © 1997 Landes Bioscience.

been used in the large-scale production of viruses for vaccines against diseases such as polio, measles, rabies, and rubella. Hybridomas have played an important role in the large-scale production of antibodies used extensively in research and medical diagnostics. The further development of tissue engineering may make cells or cell-containing tissue equivalents desired commercial products in the near future.

Physical Forces

Origin of Physical Forces

Physical forces are an important component of a cell's environment. These forces can be divided into two general categories: body and surface forces. Body forces are applied throughout the object and result from gravity and inertia. Surface forces, as their name suggests, act on the surface of an object. Some common examples of surface forces are pressure, shear stress and applied forces such as a push. As discussed below, the force that gravity exerts on a single cell is rather small. The cumulative force gravity applies on an object increases with its mass, however, so that in large clusters of cells (e.g., tissues or entire organisms), the secondary forces generated by gravity may be considerable. For example, the force gravity exerts on a single cell on the bottom of a person's foot is negligible, but the cumulative force resulting from gravity acting on other cells in the body which in turn act on that single cell may be considerable. The motion of an object, such as a blood cell moving through the vascular system, generates additional forces, including inertial forces, as well as shear stresses and pressure resulting from motion relative to the surrounding fluid. Alternatively, the motion of fluid over the surface of a stationary cell, for example an endothelial cell exposed to flowing blood, subjects the cell to shear stresses and pressure gradients. Additional forces are applied to a cell resulting from muscle contraction and adhesion of cells to the extracellular matrix.

Magnitude of Physical Forces

To estimate the relative magnitude of different physical forces acting upon a typical cell, consider the following examples. Assuming the cell is 10 μm in diameter and has a density of 1.05 g/ml, it would experience a force due to gravity of 5×10^{-7} dyn* spread over a cross sectional area of $\sim 10^{-6}$ cm², producing a stress of ~ 0.5 dyn/cm². A large fraction ($\sim 95\%$) of this gravity force is countered by a buoyancy force from the surrounding fluid.

*The force due to gravity, F, acting on a spherical object of diameter d and density ρ is $F = \dfrac{\pi d^3 \rho g}{6}$.

If this cell is subjected to an upward flow of a fluid with a velocity greater than the settling velocity of the cell in the fluid, v_t, the total shear force acting on the cell is greater than the force of gravity. Since v_t is very small, ~0.0003 cm/s,** even low-velocity fluid flow will result in shear forces greater than gravity acting on the cell. A typical shear stress due to blood flow acting upon the endothelium is 25 dyn/cm² while the force exerted by a blood pressure of 100 mm Hg is much larger, approximately 130,000 dyn/cm². The stresses resulting from muscle contraction are larger still, roughly 1,000,000 dyn/cm².***

One might erroneously conclude from the above analysis that forces due to muscle contraction and perhaps blood pressure are dominant in controlling cellular response while those due to fluid flow and gravity are insignificant. Larger forces, however, are not necessarily more important than smaller forces in determining biological response. For example, although the force that blood pressure exerts on the vessel wall is about 10,000 times larger than the force exerted by shear stress, the blood vessel can respond to each force, and both are important for proper vasoregulation. Increased shear stress results in vasodilation (chapter 2) while increased pressure results in vasoconstriction (chapter 5).

Methods of Applying Physical Forces to Cells
In Vitro

Physical forces play a prominent role in influencing physiological and pathological functions in vivo. Cell culture provides a powerful tool enabling researchers to control in vitro conditions better than in vivo conditions. But it is possible for researchers to study the effects of physical forces on cultured cells only if systems for applying the appropriate forces exist. The remainder of this chapter will briefly address the systems most commonly used to expose

---

**The terminal velocity, v_t of a sphere settling under the influence of gravity is*

$$v_t = \frac{(\rho_{sphere} - \rho_{fluid})d^2 g}{18\mu}$$

where d is the diameter of the sphere, μ is the viscosity of the fluid, and ρ_{sphere} and ρ_{fluid} are the density of the sphere and fluid, respectively.*

***Most people can comfortably curl a 10,000 g weight. Assume that most of the force is generated by the biceps and that, due to the leverage action in the arm, the actual force generated by the muscle is much greater than the force of gravity on the weight (approximately 10 times). Obviously the force of the muscle is spread over many cells. Dividing the force by 80 cm², the approximate cross-sectional area for the biceps (assuming a circular muscle with a diameter 10 cm), an average stress of 1,250,000 dyn/cm² is obtained which is about 10 times larger than the blood pressure.*

cultured cells to forces in vitro. An overview and comparison of the devices is presented in Table 1.1. Readers interested in a more detailed discussion of these systems or the derivation of equations presented here are referred to a recent review.[1]

Fluid Flow

Fluid flow can be divided into two major types, laminar and turbulent. Fluid motion in laminar fluid flow can be conceptualized as well-defined layers gliding smoothly over adjacent layers. Any transport of momentum or material between layers is by diffusion. Turbulent flow, in contrast, is characterized by random fluctuations in flow velocities superimposed on the average flow. A common example in which both laminar and turbulent flow can be observed is smoke rising from a cigarette. The smoke initially rises as a well-defined column (laminar flow). As the smoke rises, however, the flow degenerates into turbulent flow, which is observed as disordered waves of smoke. Laminar flow often is assumed to be well organized while turbulent flow is associated with complicated flow patterns, but this is not always the case. An example of a very complicated laminar flow of biological significance is that of blood flowing through bifurcation of a large artery. This laminar flow pattern exhibits vortices, flow reversals, and flow separations, all of which vary as a function of time.

In contrast to the complicated laminar flows that exist in vivo, devices used to expose cultured cells to laminar fluid flow typically are designed to apply simpler well-defined flows. For the derivation of equations describing these flows, the fluid is assumed to be an incompressible Newtonian fluid. The assumption of incompressibility is likely valid as water (and hence almost all other biological fluids) is virtually incompressible at relevant pressures. In Newtonian fluids, shear stress is linearly related to shear rate (the derivative of velocity with respect to position) by a constant of proportionality, the viscosity of the fluid. Water is almost a perfectly Newtonian fluid. Other biological fluids such as blood, however, are non-Newtonian, and thereby limit the accuracy of the predictions made by these equations.

The four major systems for subjecting cultured cells to laminar flow are the parallel-disk viscometer, the cone-and-plate viscometer, the parallel-plate flow chamber and the radial flow chamber. As their names suggest, the first two systems were originally adapted from devices used to study the rheological properties of fluids. What follows is a brief description of each system along with a discussion of its major advantages and disadvantages as well as some experimental applications.

Parallel-Disk Viscometer

The parallel-disk system consists of a stationary plate and a rotating disk (Fig. 1.1). The surfaces of the disk and plate are parallel and are separated by a distance, h. The volume separating the two surfaces is filled with liquid. The cells to be subjected to mechanical forces are either cultured adherent to the stationary plate or are suspended in the medium. The rotation of the upper disk, with an angular velocity ω, causes the fluid to move in a circumferential direction with a velocity v_θ. The velocity varies as a function of the distance from the axis of rotation, r, and the distance from the stationary disk, z, as described by the following equation.

$$v_\theta = \frac{r\omega z}{h}$$

The shear stress, τ, exerted by the moving fluid will also vary as a function of position but only on radial distance, r, with a maximum at the edge of the disk and zero at the center.

$$\tau = \mu \frac{r\omega}{h}$$

This distribution of shear stresses may be either beneficial or detrimental, depending on the desired application. For example, this distribution would be beneficial in cases where researchers wanted to analyze the morphology of adhered cells subjected to a range of shear stresses, since it would be possible to observe cells subjected to a wide array of shear stresses in a single experiment. This system would not be appropriate in cases where it is desired to expose cells to a uniform shear stress; in these cases, other systems such as the cone-and-plate viscometer or parallel-plate flow chamber should be used.

A potential concern when using either a parallel-disk or cone-and-plate viscometer is the presence of secondary flows. In each device, the fluid near the rotating surface will have a higher velocity than the fluid near the stationary surface. Due to its higher velocity and the fact that it is traveling in a circular path, the fluid at the top of the device near the moving surface experiences a larger centrifugal force. This imbalance in centrifugal forces results in secondary flow patterns of fluid flowing outwardly near the top surface and inwardly along the bottom surface. For a more complete discussion of secondary flows and the conditions under which they may be significant, see the review by Tran-Son-Tay.[1]

Cone-and-Plate Viscometer

Unlike in a parallel-disk viscometer, the distance h between the two surfaces in a cone-and-plate viscometer increases with the distance from the

Table 1.1. A summary of devices and methods used to study the effects of physical forces on animal cells in culture. For more complete description of each system, see text.

Name of device/ method of developing force	Primary force	Spatial distribution of forces	Fluid flow in and out of device?	Comments
Parallel-disk viscometer	Shear stress	Shear stress directly proportional to radial position	No	Possible presence of secondary flows
Cone-and-plate viscometer	Shear stress	Ideally shear stress is not a function of position	No	Possible presence of secondary flows
Parallel-plate flow chamber	Shear stress	Shear stress is not a function of position; pressure varies linearly as a function of position, though this variation is often ignored as it is very small	Yes	Possible presence of not fully developed laminar flow where fluid enters device
Radial flow chamber	Shear stress	Shear stress varies as a function of position; greatest near the center of the device	Yes	Presence of not fully developed laminar flow where fluid enters device

Uniaxial stretch	Tensional Stress	Greatest at the center of the device	No	May also generate shear stress due to motion of cells relative to fluid
Biaxial stretch	Tensional Stress	Greatest at the edge of the device	No	May also generate shear stress due to motion of cells relative to fluid
Compression of gas phase by addition of an inert gas such as He	Pressure	Uniform	No	Change in concentration of dissolved gas due to nonideal behavior
Direct compression of liquid phase	Pressure	Uniform	No	No gas phase present

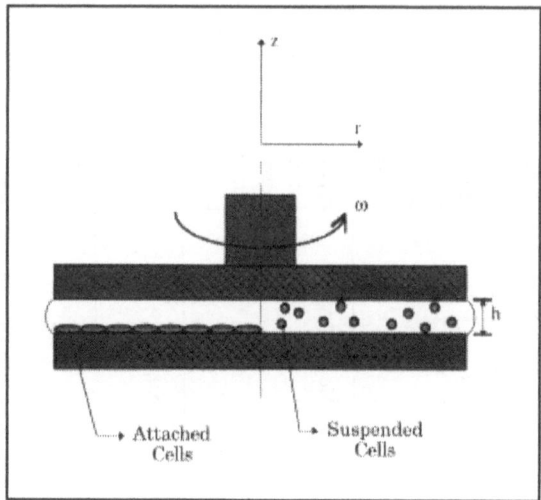

Fig. 1.1. A schematic diagram of a parallel-disk viscometer consisting of a stationary and a rotating disk.

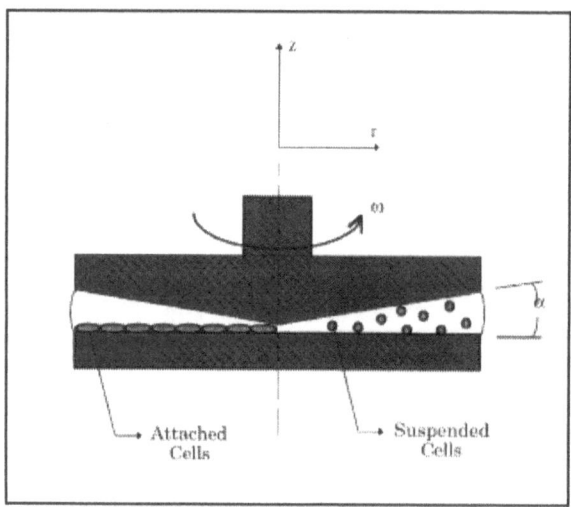

Fig. 1.2. A schematic diagram of a cone-and-plate viscometer.

center, r (Fig. 1.2). This geometric difference provides constant shear stress regardless of position. The simplest analysis of this system takes advantage of the fact that as the angle between the cone and the plate, α, is small, the flow in any local region can be approximated as flow between two parallel surfaces.

As before,

$$v_\theta = \frac{r\omega z}{h}$$

recognizing that h is not a constant but is now a function of r, namely, h = r sin a we obtain

$$v_\theta = \frac{r\omega z}{r\sin\alpha} = \frac{\omega z}{\sin\alpha} \approx \frac{\omega z}{\alpha}$$

An interesting variation of the cone-and-plate viscometer is the rheoscope. In the rheoscope, the cone and plate rotate in opposite directions with equal angular velocity. This provides an area equidistant between the cone and the plate in which the fluid does not move relative to the room. This allows for observation of the deformation of suspended cells without high-speed cinematography. The shear stress at any point in a rheoscope is twice that of a corresponding cone-and-plate viscometer with the same angular velocity, a characteristic due to the opposing rotation of the surfaces.

Parallel-Plate Flow Chamber

There are two major types of flow chambers, parallel-plate and radial flow chambers. Unlike the systems described above in which fluid motion is caused by the motion of at least one surface in the device, fluid flow in a flow chamber is caused by an imposed pressure gradient. In these devices, fluid is continuously introduced at one side of the chamber while simultaneously removed from the other side. This difference may be a limitation in some cases; for instance, an attempt to expose suspended cells to fluid flow for an extended period would result in these cells being swept out of the device. Alternatively, this characteristic may be beneficial as it permits the convenient addition and removal of medium.

The most commonly used flow chamber is the parallel-plate flow chamber (Fig. 1.3). The pressure gradients driving the flow can be established by either a hydrostatic head or by some form of pump; each has its advantages. A hydrostatic head provides extremely even flow, whereas a pump permits rapid variations in flow and shear stress. These variations result from the incompressibility of the fluid, and any changes in flow rate through the pump must be matched by a rapid increase in flow rate through the flow chamber.

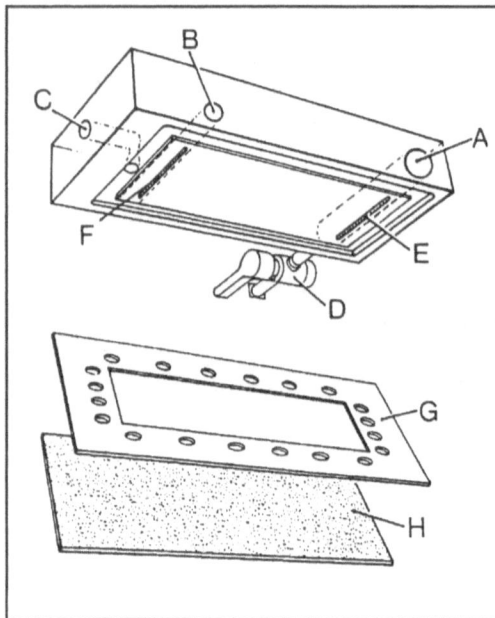

Fig. 1.3. Parallel plate flow chamber. The polycarbonate plate, the gasket (G), and the glass slide (H) with the attached cells are held together by a vacuum (C), forming a channel of parallel plate geometry. Medium enters at port entry (A), through slit (E), into the channel, and exits through slit (F), and exit port (B). Entry port (A) also serves as a trap for bubbles, which can be removed through valve (D). With permission from Frangos JA, McIntire LV, Eskin SG. Shear stress induced stimulation of mammalian cell metabolism. Biotech Bioeng 1988; 32:1053-60.

For a flow rate Q through a chamber of breadth b and height h, the wall shear stress is independent of position and is predicted by the following equation:

$$\tau_{wall} = \frac{6Q\mu}{bh^2}$$

Frequently, cells are cultured on a glass microscope slide and then attached to the chamber to be exposed to flow. The glass slide is sealed to the flow chamber with a gasket and held firmly in place by a series of clamps or by a uniform vacuum applied around the perimeter of the chamber.

Radial Flow Chamber

In a radial flow chamber fluid is introduced in the center, moves out radially, and exists at the outer edge (Fig. 1.4). The wall shear stress is maximal at the center of the chamber and decreases with distance from the center. The wall shear stress is described by the following equation.

$$\tau_{wall} = \frac{6Q\mu}{2\pi rh^2}$$

As with the parallel-disk viscometer, the radial dependence on shear stress limits the application of this device, but in specific instances this may be advantageous. For instance, the radial flow chamber has been used to study the effect of shear stress on the adhesion and rolling of neutrophils on an

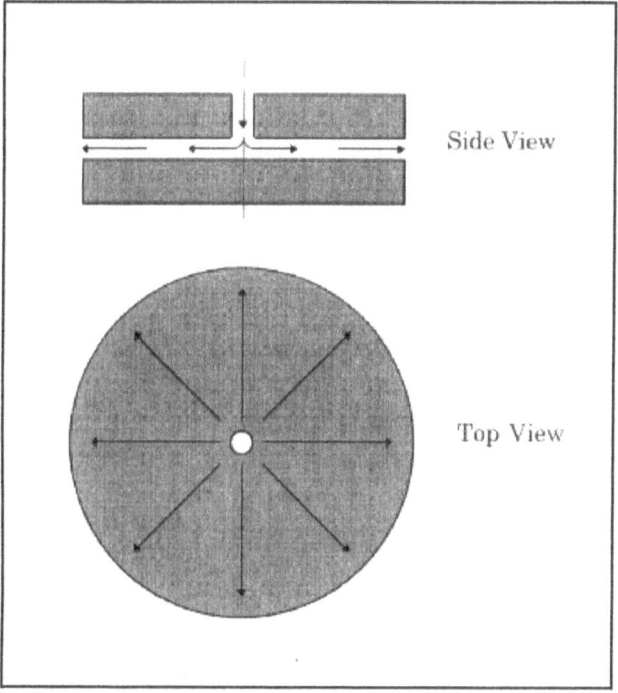

Fig. 1.4. A schematic diagram of a radial flow chamber.

endothelial cell monolayer. Since the shear stress varies from a very high level at the center to a much smaller level at the edge, it is possible to observe a wide range of shear stresses in one system.

Turbulent Systems

Conceivably, each of the four systems discussed above could be used to subject cultured cells to turbulent flow. Davies et al[2] demonstrate that endothelial cells exposed to turbulent flow in a cone-and-plate viscometer have a more rapid turnover than similar cells subjected to laminar flow with the same average shear stress or subjected to no flow conditions. Normally, however, these devices are operated to generate only laminar flow.

Typically, specific devices are not designed for exposing cells to turbulent flow; instead, pre-existing cell culture systems such as spinner flasks or stirred-tank bioreactors are used. Normally the turbulent flow that results is only crudely characterized, and often only with parameters having little descriptive value. The limitation of such information is apparent when one attempts to compare data from different reports. The lack of specific systems to apply well characterized turbulent flow makes the interpretation and comparison of data extremely difficult.

Stretch

Devices to apply stretch are used by researchers to approximate the strain experienced by cells in vivo. A common method of stretching cells involves plating the cells on a flexible surface, such as a plastic membrane, and cyclically or statically stretching the membrane with fixed amplitude and duration. A problem with cell deformation devices of this nature lies in the difficulty of determining whether the cells are fully adherent to the flexible membrane. Cells that have not fully adhered to the plate do not experience the same amount of stretching as those completely affixed, and the subsequent data may be misleading. Nevertheless, this method of applying stretch to cell cultures is the method cited commonly in the literature. Uniaxial and biaxial stretch, two variations of this method, are described below.

Uniaxial Stretch

The earliest attempts to subject animal cell culture to strain used uniaxial stretching devices, those in which the boundaries of the membrane are stretched. Of these, the first type to be used probably was a flexible culture dish affixed to two blocks that could be forced apart by turning a screw (Fig. 1.5). This device provides fairly uniform stretching, although stretch is greatest at the center. It perhaps is most useful for experiments investigating static stretch, because devices of this design sometimes can permanently alter the shape of the culture dish.

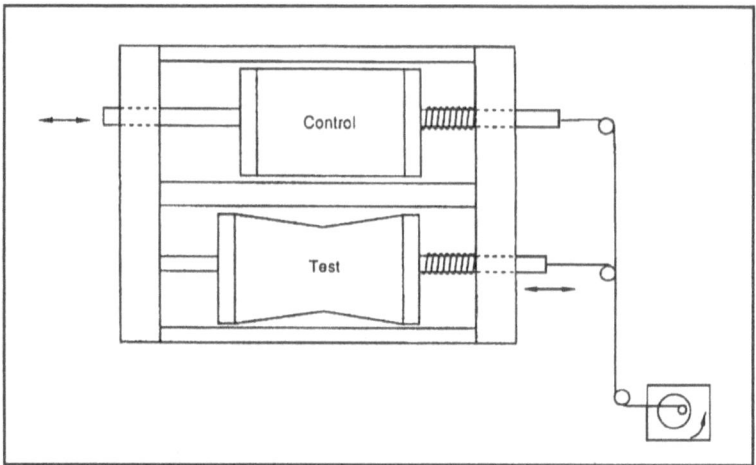

Fig. 1.5. A schematic diagram of a uniaxial stretch device. Reprinted with permission from Academic Press © 1993. Physical Forces and the Mammalian Cell. Frangos JA, ed. San Diego: Academic Press, 1993.

Biaxial Stretch

Biaxial stretch mechanisms differ from uniaxial devices in that the outer edges of the plate are fixed and stretch is applied by cyclic pressure above or below a round culture dish. The area of greatest stretch in these devices is near the outer edge (Fig. 1.6). As with uniaxial devices, adhesion of cells to the culture dish can be a variable that is difficult to determine. This type of stretching device is desirable for conducting cyclic stretching trials, because boundaries of the culture dish are maintained and the shape of the dish usually is not permanently altered.

Pressure

Many initial attempts to subject cells to elevated pressure utilized compressed gas. A major drawback of this method, however, is that the solubility of a gas in liquid increases as its pressure increases. This relationship can be approximated by Henry's Law, which states that the solubility of a gas increases linearly with its partial pressure with a proportionality constant unique to each gas. As the concentration of some dissolved gases (notably oxygen and carbon dioxide) affects the metabolic activity of cells, it is difficult to determine whether the effects observed in this system result from the pressure or from the change in gas concentration.

Two methods of circumventing this problem have been developed. The first is to increase pressure by the addition of an inert gas such as helium while maintaining the partial pressure of the other gases. Doing this increases the concentration of the dissolved helium, which presumably does not have a biological effect, while the concentration of other dissolved gases remains the same. Alternatively, researchers have increased pressure by directly compressing the fluid in the absence of a gas phase.

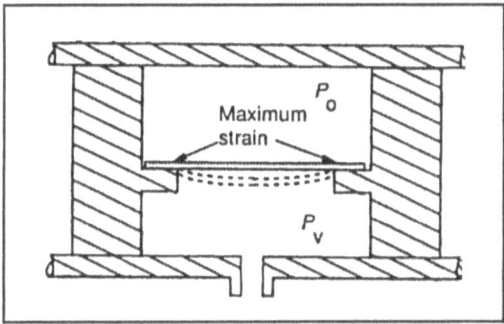

Fig. 1.6. A schematic diagram of a biaxial stretch device. Reprinted with permission from Academic Press © 1993. Physical Forces and the Mammalian Cell. Frangos JA, ed. San Diego: Academic Press, 1993.

Summary

Animal cell culture is an important tool in the study of complex biological systems. Transferring cells from their original environment in vivo to a relatively simple in vitro environment facilitates the isolation and control of the variables to be tested. But with this benefit comes a loss of representativeness, the degree to which the setting is analogous to the in vivo environment.

Physical forces are an important component of a cell's environment and need to be maintained in culture as faithfully as possible to construct accurate models of in vivo systems. The physical forces to which cultured cells are most commonly exposed are fluid flow, pressure, and stretch. The magnitude of physical forces acting upon a typical cell can vary from <0.5 dyn/cm^2 to 10^6 dyn/cm^2. Larger forces, however, are not necessarily more important than smaller forces in determining biological response. Devices commonly used to subject cultured cells to flow, pressure and stretch are summarized in Table 1.1.

References

1. Tran-Son-Tay R. Techniques for studying the effects of physical forces on mammalian cells and measuring cell mechanical properties. In: Frangos JA, ed. Physical Forces and the Mammalian Cell. San Diego: Academic Press, 1993:1-60.
2. Davies PF et al. Turbulent fluid shear stress induces vascular endothelial cell turnover in vitro. Proc Natl Acad Sci USA 1986; 83:2114-7.

CHAPTER 2

Endothelial Cells

A therosclerosis is the leading cause of death in the United States and most
Western countries. Despite intense research efforts, much is still unknown
about the etiology of the disease. While many clinical and biochemical factors have been linked to atherosclerosis, none of these factors can account for
the extreme localization of this disease within the vascular system.[1] The carotid bifurcation, the coronary arteries, the distal abdominal aorta and the
major arteries of the leg are susceptible, while most other medium-sized and
large arteries, including the pulmonary and mesenteric arteries and the vessels of the arm, remain disease-free.[2] Even within these specific vessels, atherosclerosis is not uniformly distributed, but diseased and healthy regions of
the vessel are often separated by less than 1 cm. Hemodynamic forces likely
account for this localization as they can vary greatly over very small distances.
Attempts to correlate lesion development with hemodynamic forces have
demonstrated that intimal thickening co-localizes with region with oscillatory
and low mean wall shear stress (Figs. 2.1 and 2.2).

Endothelium Structure and Function

Possibly due to its involvement in vascular disease and its constant exposure to a variety of mechanical forces in vivo, the endothelial cell is likely the
cell type whose responses to physical forces have been studied most extensively. In order to understand endothelial cell responses to physical forces, it
is important to first understand the cell's normal function. Therefore, a brief
description of endothelial cell biology is provided below. For a more extensive discussion of endothelial biology, see the review by Boeynaems and
Pirotton.[3]

Antithrombogenic

When blood is exposed to most surfaces, it clots as the result of a complex
cascade of biochemical events. The resulting clot, or thrombus, obstructs
blood flow. This obstruction may sometimes perform a useful function, such
as limiting internal or external bleeding. If a thrombus forms within a blood

Mechanical Forces: Their Effects on Cells and Tissues, by Keith J. Gooch and
Christopher J. Tennant. © 1997 Landes Bioscience.

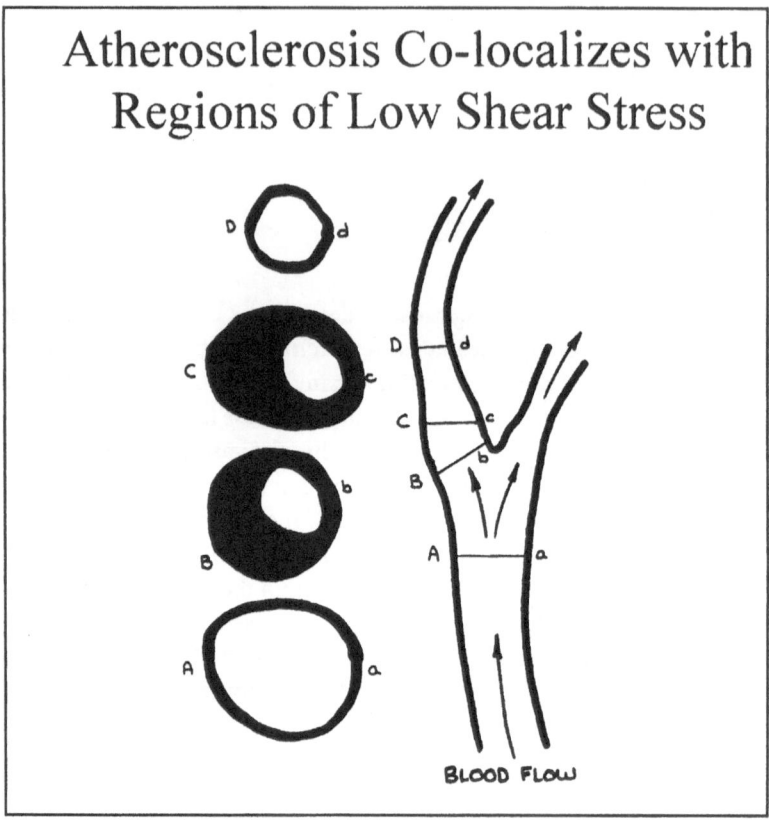

Fig. 2.1A. The spatial distribution of atherosclerotic lesions along the carotid bifurcation (Panel A). Upstream (cross section A-a) and downstream (cross section D-d) of the bifurcation are relatively disease free while the areas along the bifurcation (cross sections B-b and C-c) are partially occluded. Wall thickening at the bifurcation is asymmetrical, with the majority of the thickening occurring in the region distal to the flow divider. Adapted from Glagov S, Weisenberg E, Giddens DP et al. Hemodynamic and atherosclerosis: Insights and perspectives gained from studies of human arteries. Arch Pathol Lab Med 1988; 112:1018-31.

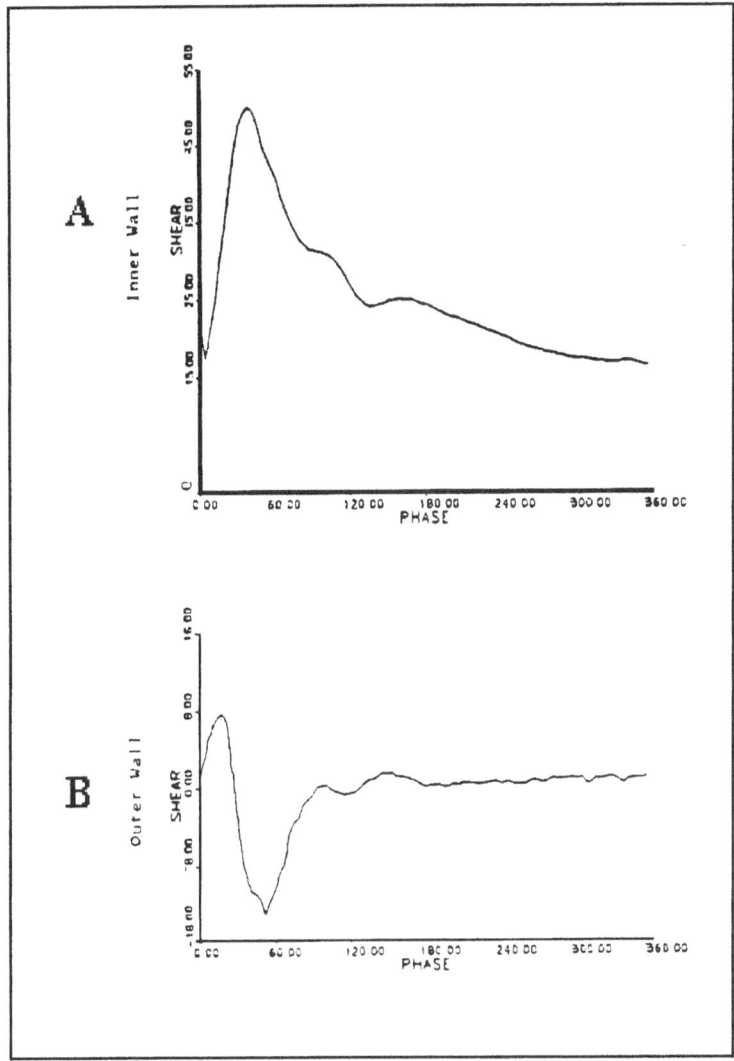

Fig. 2.1B. Pulsatile shear stress of the human carotid bifurcation in dyn/cm² for 360° of the cardiac cycle at: the inner wall flow divider (a); and the posterior outer wall (b). Shear stress at the inner wall was found to be unidirectional and high throughout the entire cardiac cycle. However, shear stress at the outer wall was found to oscillate in direction with a minimum value of -13 dyn/cm². With permission from Sumpio BE. Hemodynamic Forces and Vascular Cells. Austin: R.G. Landes Company, 1993.

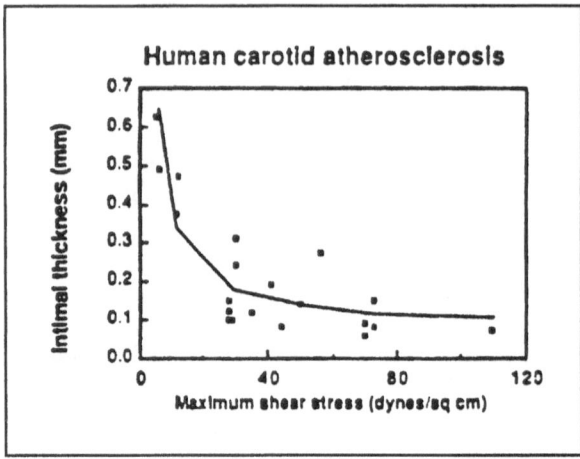

Fig. 2.2. Correlation of intimal thickening with wall shear stress for various locations along a human carotid bifurcation. Intimal thickening is much more pronounced in regions subjected to less than 10 dyn/cm². In: Sumpio BE, ed. Hemodynamic Forces and Vascular Cells. Austin: R.G. Landes Company, 1993.

vessel and obstructs blood flow, however, catastrophic results such as stroke, heart attack, or gangrene may follow.

To prevent unwanted clotting, a confluent monolayer of endothelial cells lines the cardiovascular system, forming a nonthrombogenic surface. The endothelium is nonthrombogenic due to the anticoagulant activity of the surface of, and products released by, endothelial cells. Heparin sulfate proteoglycans, which are present on the surface of endothelial cells, accelerate the neutralization of thrombin[4] while the integral protein thrombomodulin binds thrombin, modifying its substrate specificity. Endothelial cells promote fibrinolytic activity by providing a location for the assembly of the fibrinolytic system and releasing tissue (t-PA) and urokinase (u-PA) plasminogen activators.[5] Prostacyclin (PGI_2)[6] and nitric oxide (NO)[7,8] are released by endothelial cells and inhibit platelet activation by increasing the intracellular concentration of cAMP[9] and cGMP,[10] respectively. As prostacyclin and NO work through distinct second messenger systems, they act synergistically to inhibit platelet aggregation.[11]

Vasoregulation

PGI_2 and NO also increase intracellular cAMP and cGMP concentrations in vascular smooth muscle cells, leading to smooth muscle cell relaxation by distinct mechanisms. Binding of PGI_2 to receptors on endothelial cells stimulates a cascade of events leading ultimately to smooth muscle relaxation through a cAMP-dependent mechanism.[6] Though prostacyclin regulates the tone of some vascular beds,[6] NO is the major endothelium derived relaxing factor.[7,8] NO induces smooth muscle cell relaxation through both cGMP-dependent[12] and -independent mechanisms.[13] This smooth muscle relaxation permits vasodilation and increased blood flow. Alternatively, endothelial cells may produce the potent vasoconstrictor endothelin-1 (ET-1). ET-1 binds to ET_A receptors on smooth muscle cells, which leads to a rise in intracellular Ca^{2+}, resulting in contraction.

Control of Smooth Muscle Proliferation

In addition to acute control of contraction, endothelial cells exercise control of smooth muscle function by influencing proliferation. Endothelial cells produce platelet-derived growth factor (PDGF)[14,15] and ET-1,[16,17] both smooth muscle mitogens, as well as a heparin-like inhibitor of smooth muscle proliferation.[18] NO and PGI_2 both have been reported to inhibit smooth muscle cell proliferation,[19] though the concentration of NO required may exceed that expected in physiological circumstances.[20,21]

Control of Transvessel Transport

In addition to regulating the flow through blood vessels, the endothelium plays an active role in the regulation of the transport of water, solutes, macromolecules and cells across the vessel wall. Pressure and concentration gradients drive water and solutes across the blood vessel wall, while the endothelium acts as a major resistance to transport. Two different transendothelial pathways can be utilized for transport: diffusion and convection through intercellular[22] and perhaps intracellular junctions,[23] or vesicle-mediated nondiffusive transport of receptor-bound or fluid-phase material via transcytosis.[24]

The permeability of the endothelium varies along the vascular tree and according to the location of the vascular bed. In addition to spatial distributions, many pro-inflammatory compounds such as histamine, serotonin and thrombin increase permeability,[25,26] although increases in permeability appear to be secondary to changes in cellular contraction. The reorganization of F-actin microfilaments and phosphorylation of myosin light chains both have been suggested as mechanisms regulating cellular contraction.[27,28]

Endothelial Cell Responses to Physical Forces

As the interface between the flowing blood and the rest of the blood vessel, all endothelial cells in vivo are constantly exposed to both tangential shear stress and normal pressure. Cardiac and arterial endothelial cells are exposed to cyclic strain accompanying the cardiac cycle, while venous endothelial cells are strained by venous pumping, a process in which a system of one-way valves and muscular contraction help to carry blood back to the heart. Clearly, endothelial cells exist in a mechanically active environment in vivo. In light of the prevalence of mechanical forces in the environment of these cells, it is not surprising that they are able to respond to the forces. What is surprising is the large number of responses initiated by physical forces.

Endothelial Cell Responses to Fluid Flow

Fluid flow, probably acting through shear stress, regulates many physiologically significant endothelial functions. In fact, it seems few functions of the endothelium studied so far are not modified by flow. The response of endothelial cells to fluid flow has been comprehensively reviewed[29-31] and thus a detailed review will not be presented here. Table 2.1, condensed and adapted from an excellent review by Davies,[29] lists some of the responses that have been studied. Potential causal relationships between observed flow-induced responses are presented in Figure 2.3. Only a few of these responses (those of historical significance, of exceptional physiological importance or those that illustrate a specific point) will be discussed here.

Second-messenger systems

As would be suggested by the large number of biochemical responses triggered by fluid flow, many second-messenger systems are activated by flow. As is shown in Table 2.1, flow stimulates ion-channel activation, G-protein turnover, transient increases in intracellular Ca^{2+}, phosphorylation of regulator proteins and the production of the second messengers cAMP, cGMP, inositol 1,4,5-triphosphate (IP_3) and diacylglycerol (DAG). It is somewhat surprising that with such a large number of signal transduction mechanisms stimulated, the result is not a chaotic disruption of the cellular metabolism. Instead, a well-coordinated response often is observed.

Antithrombogenic

Fluid flow stimulates endothelial cells to increase their production of the antithrombogenic compounds t-PA,[54,56] as well as PGI_2,[52,53] and NO[32,70] as will be discussed in more detail in the following section on vasoregulation. Human umbilical vein endothelial cells (HUVEC) exposed to steady flow of a shear stress of 25 dyn/cm[2] secreted t-PA at a nearly constant elevated rate after a 6-h lag. The rates of secretion were 2.1 and 3.1 times the basal rate of 0.168 ng/10[6] cells/h for cells subjected to 15 and 25 dyn/cm[2], while cells exposed to 4 dyn/cm[2] did not exhibit increased production. In the same study, the production of plasminogen activator inhibitor (PAI-1) was unaffected by

Table 2.1. Flow-induced responses by cultured endothelial cells.

Effect	Force	Cell Type & Time	Potential Significance	Refs.
Ionic				
K^+ channel activation; hyperpolarization (whole cell recording)	laminar shear stress (LSS); 0.2-16.5 dyn/cm²	BAEC seconds	increased driving force for calcium entry	32-34
Hyperpolarization (membrane potential-sensitive fluorescent dyes)	LSS; 10-120 dyn/cm²	BPAEC steady state at 60 s	increased driving force for calcium entry	35
Rb^+ efflux stimulated	LSS; 1-10 dyn/cm²	PAEC permeability	graded transient increase of K^+	36
Activation of non-selective cation channels (membrane patch)	suction (pressure, stretch) 10-20 mm Hg	PAEC ms	endothelial stretch-activated channels	37
Intracellular Ca^{2+} rise (fluo-3)	mechanical poking and dimpling	HUVEC seconds	activation of calcium-dependent pathways	38
Decrease of intracellular pH	LSS; 0.5-13.4 dyn/cm²	BAEC	modulation of ionic balance and enzyme activities	39
Intracellular Ca^{2+} rise; Ca^{2+} oscillations	LSS; 0.2-4.0 dyn/cm²	BAEC 15-40 s	activation of calcium-dependent pathways	40-43

Table 2.1. (continued)

Effect	Force	Cell Type & Time	Potential Significance	Refs.
Release of compounds				
Large increase in release of NO	LSS; 8 dyn/cm^2	BAEC seconds	flow-mediated vasorelaxation	32,44-46
Release of ATP, acetylcholine, endothelin and substance P	Flow through microcarrier bed	HUVEC seconds		47-50
Sustained PGI$_2$ release	LSS; 0-9 and 14.0 dyn/cm^2	HUVEC 2 min	flow-mediated vasorelaxation	51-53
Release of t-PA	LSS; 15 and 25 dyn/cm^2	HUVEC 5 hours	enhancement of fibrinolytic activity	54
Histamine release and histamine decarboxylase activity stimulated	oscillatory LSS; range 1.6-8.2 dyn/cm^2	BAEC >6 hours	modulation of endothelial permeability barrier	55
Gene Transcription and Translation				
t-PA mRNA expression	LSS; 15 and 25 dyn/cm^2	HUVEC 5 hours	enhancement of fibrinolytic activity	56
Activation of NFκB	LSS; 10 dyn/cm^2	BAEC 20 min	transcription factor activation	57
Induction of c-myc, jun	LSS; 10 dyn/cm^2	BAEC 30 min	transcription factors	58

Downregulation of VCAM-1 expression	LSS; 0-7.2 dyn/cm²	mouse lymph node endothelial cells >1 hour	preferential leukocyte adhesion at low shear stress	59
Tenfold increase in PDGF$_A$ mRNA; peaks at 1.5-2 hours	LSS; 0-51 dyn/cm²	HUVEC, BAEC >1 hour	regulation of cellular growth	60-62
2- to 3-fold increase of PDGF$_B$ mRNA followed by 4-fold decrease by 9 hours; PKC dependence controversial	LSS; 10-36 dyn/cm² steady, pulsatile, turbulent	HUVEC, BAEC 1-9 hours	regulation of cellular growth	60,61,63
bFGF mRNA stimulated 1.5- to 5-fold	LSS; 15 and 36 dyn/cm²	BAEC 0.5-9 hours (peak at 6 hours)	regulation of cellular growth	62
NO synthase mRNA increased	Fluid flow	BAEC 3 hours	vasodilation	64
Heat shock protein-70 mRNA increased 2- to 4-fold	LSS; bidirectional	BAEC 4 hours		65
Decreased thrombomodulin mRNA & protein at 15 and 36 dyn/cm²	LSS; 4, 15, 36 dyn/cm²	BAEC 9 hours	protective role against thrombosis in regions of low shear stress	66
Signal Transduction Mechanisms				
Transient elevation of IP$_3$ biphasic	LSS; 30 and 60 dyn/cm²	BAEC HUVEC >15-30 s cyclic stretch (BAEC)	second messengers regulating release of calcium from intracellular stores major peak at 5 min	67, 68

Table 2.1. (continued)

Effect	Force	Cell Type & Time	Potential Significance	Refs.
cGMP increased 3-fold via a NO-dependent mechanism	LSS; 0–40 dyn/cm²	BAEC 60 s	vasoregulation of ET-1 release	69-71
Transient elevation of IP₃	cyclic strain, 1 Hz; deformation, 24%	HUVEC	second messengers regulating release of calcium from intracellular stores	72, 73
Activation of adenylyl cyclase	cyclic stretching; osmotic swelling	BAEC, HUVEC minutes	second messenger	74
Morphology and cytoskeletal rearrangement				
Cell alignment in direction of flow; function of time and magnitude of shear stress	LSS; >5 dyn/cm²	all types > 6 hours	minimizes drag on cell	75-79
F-actin cytoskeletal and fibronectin rearrangement	LSS; >5 dyn/cm² and in vivo	all types > 6 hours	associated with cell realignment	75,80-85
Differential cell shape and alignment responses; corresponding F-actin changes	LSS; pulsatile 1 Hz; sinusoidal flows of various patterns up to 60 dyn/cm²	BAEC >6 hours		86-88
Mechanical stiffness of cell surface proportional to extent of realignment to flow	LSS; 10-85 dyn/cm²	BAEC 24 hours	decrease deformability of sub-plasma membrane cortical complex	89

Proliferation

Stimulation of cellular proliferation in quiescent monolayer	turbulent flow; average shear stress 1.5–15.0 dyn/cm^2	BAEC >3 hours	loss of contact, inhibition of growth by disturbed flow	90
Inhibition of cellular proliferation in subconfluent monolayers	LSS; 7–90 dyn/cm^2	BAEC 24 hours	inhibition of endothelium regrowth	91
Regional cell cycle stimulation in confluent monolayer	disturbed laminar flow (flow separation, vortex, reattachment) 0–10 dyn/cm^2	BAEC 12 hours	steep shear gradients stimulate cell turnover; focal hemodynamic effects	92

LSS, laminar shear stress; BAEC, bovine aorta endothelial cells; HUVEC, human umbilical vein endothelial cells; PAEC, porcine aortic endothelial cells. Adapted from a more extensive table presented in an excellent review by Davies.[29]

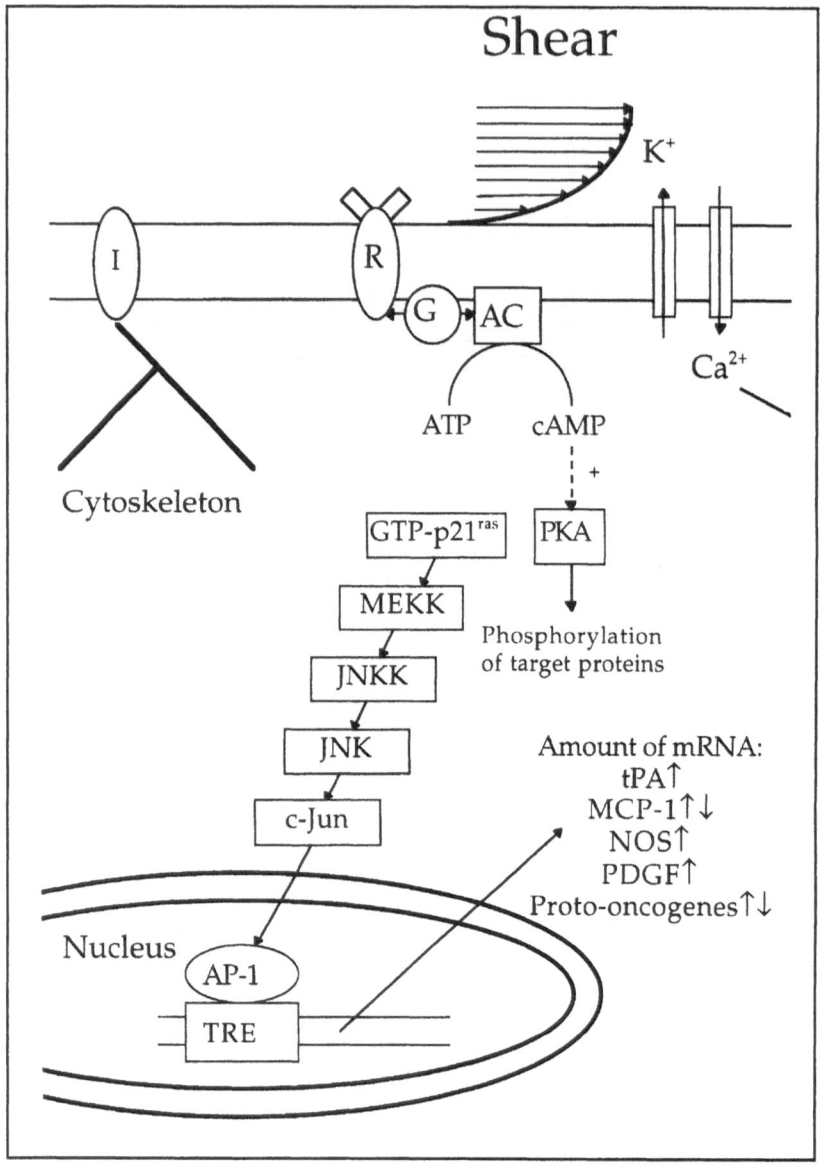

Fig. 2.3. (see legend, opposite page).

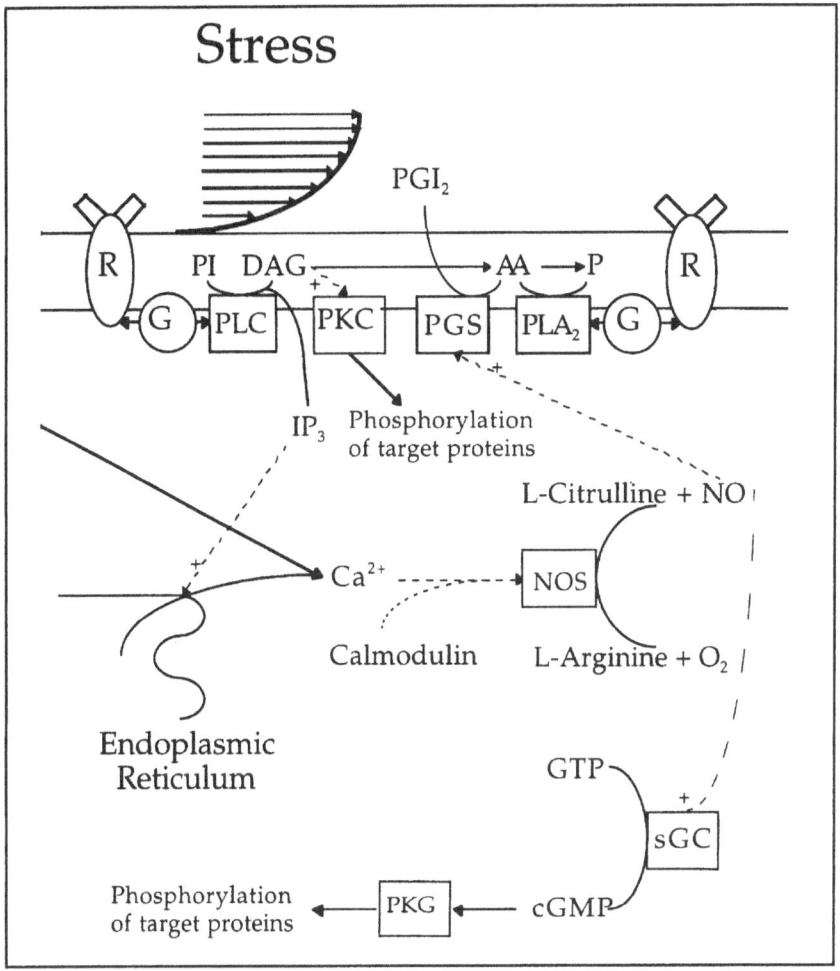

Fig. 2.3. Interactions between some of the effects of shear stress on endothelial cells. Abbreviations used are as follows: AA, arachidonic acid; AC, adenylate cyclase; AP-1, transcription factor AP-1; ATP and GTP, adenosine and guanosine 5-triphosphate, respectively; cAMP and cGMP, adenosine and guanosine 3,5-cyclic monophosphate, respectively; DAG, 1,2-diacylglycerol; G, G protein; I, integrin; IP_3, inositol 1,4,5-triphosphate; JNK, c-jun aminoterminal kinase; JNKK, c-jun aminoterminal kinase kinase; MCP-1, monocyte chemotactant protein-1; MEKK, mitogen associated protein kinase kinase; NO, nitric oxide; NOS, nitric oxide synthase; P, phospholipid; PDGF, platelet-derived growth factor; PGI_2, prostacyclin; PI, phosphatidylinositol; PGS, prostaglandin synthase; PKA, protein kinase A; PKC, protein kinase C; PKG, protein kinase G; PLA_2, phospholipase A_2; PLC, phospholipase C; sGC, soluble guanylate cyclase; R, receptor; tPA, tissue-type plasminogen activator; TRE, 12-O-tetradecanoyl-phorbol-13-acetate responsive element. © Keith Gooch, used with permission.

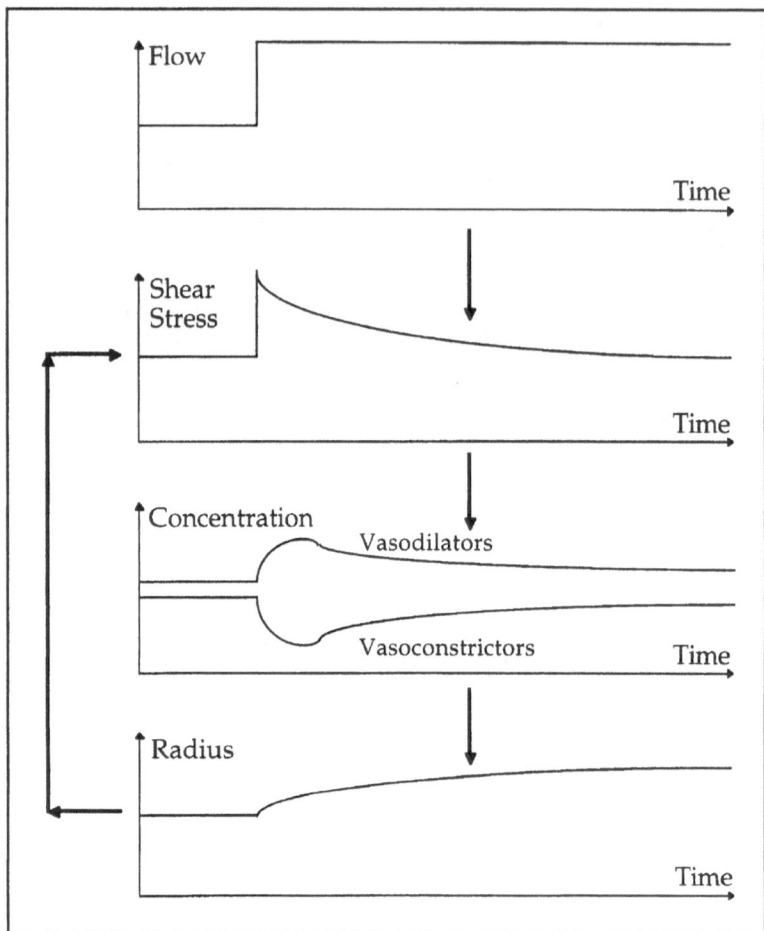

Fig. 2.4. Vasoregulation provides a vivid example of how mechanical stimulation initiates a well-coordinated response by the cells. A graphical illustration of vasoregulation in response to a step-change increase in blood flow. The increased blood flow initially results in an increase in wall shear stress and a subsequent increase the concentration of vasodilators (e.g., nitric oxide) and a concurrent decrease in the concentration of vasoconstrictors (e.g., endothelin-1). The net effect of these vasoactive compounds is smooth muscle cell relaxation and an increase in vessel radius. As the radius increases, the wall shear stress decreases, thereby completing a feedback control loop. Note that this figure is intended to illustrate causal rationships between the above variables, not to provide quantitative information on their magnitudes or rates of change.

the flow regimes studied, suggesting that flow enhances the endothelium's antithrombogenic property by stimulating production of a fibrolytic compound but not of its corresponding inhibitor.

Vasoregulation

In vivo and ex vivo studies[32] have demonstrated that increasing the flow rate of blood or other perfusing medium results in acute vessel dilation. If an increased blood flow rate in vivo persists over a period of weeks, blood vessel diameter increases.[93,94] Both responses are dependent on a functional endothelium. As fluid flows through a vessel, it exerts a tangential frictional force, shear stress (τ), on the vessel wall. For Newtonian fluids, this force is described

by the equation $\tau = -\mu \dfrac{dv}{dr}$

so that the magnitude of the force is proportional to the velocity gradient at the wall (dv/dr) and the viscosity of the fluid (μ). Increasing fluid viscosity, while keeping the flow rate constant and hence the velocity gradient at the wall, increases flow-induced dilation,[95] suggesting that vasodilation is caused by the shear stress exerted by the fluid rather than by the motion of the fluid itself. These observations suggest a regulatory feedback control system in which increases in flow and wall shear stress result in vasodilation (Fig. 2.4). Wall shear stress decreases as the diameter of the vessel increases, returning the wall shear stress to its set point value. Subsequent studies have indicated that shear stress stimulates cultured endothelial cells to increase their release of the vasodilators PGI$_2$[52,53] and NO[32,70] while inhibiting the basal release of vasoconstrictor ET-1.[96] These results indicate that multiple vasoactive compounds are involved in vasodilation and that the process is well coordinated, with an increase in the release of vasodilators and a decrease in the release of the vasoconstrictor.

PGI$_2$

When either bovine aortic endothelial cells[53] or HUVEC[52] are exposed to laminar fluid flow, PGI$_2$ production increases as monitored by the accumulation of 6-keto-prostaglandin F$_1\alpha$, the stable breakdown product of PGI$_2$. Upon the onset of flow, a burst of PGI$_2$ production is followed by prolonged elevation and sustained production. The magnitude of the initial burst of production is independent of the applied shear stress (Fig. 2.5, Panel A); however, the rate of sustained production is dependent on shear stress (Fig. 2.5, Panel B).

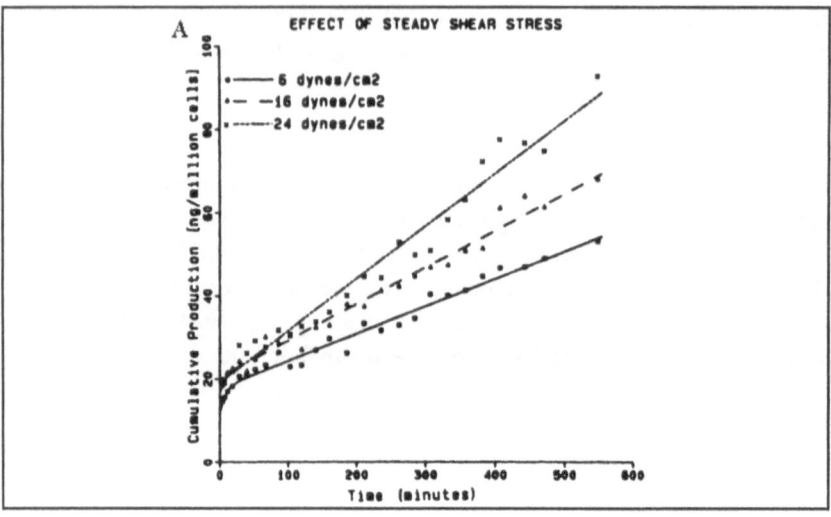

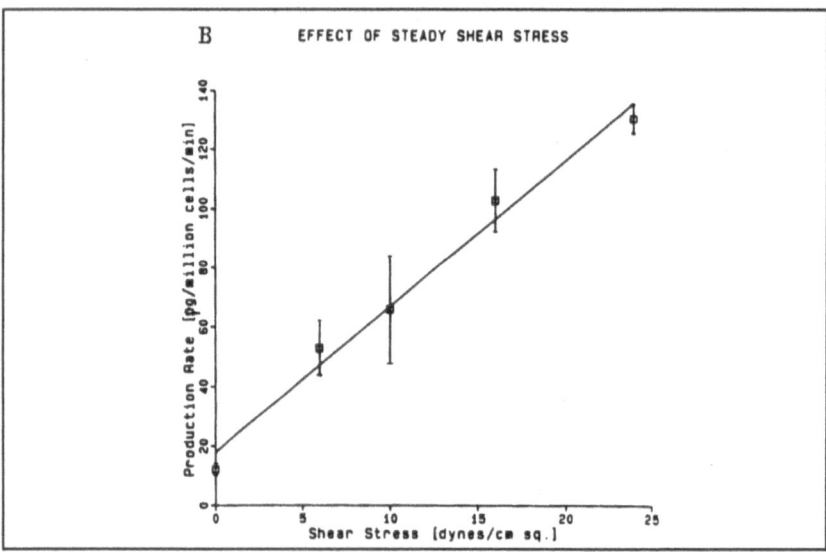

Fig. 2.5A,B. Production of 6-keto-prostaglandin F$_1\alpha$, the stable breakdown product of PGI$_2$, by HUVEC at different levels of shear stress. At the onset of flow, there is a burst in 6-keto-prostaglandin F$_1\alpha$ production followed by sustained steady production for almost 10 h (Panel A). The increase in sustained rate of synthesis is proportional to the magnitude of the shear stress applied (Panel B). With permission from Frangos JA, McIntire LV, Eskin SG. Biotechnol Bioeng 1988; 32:1053-60.

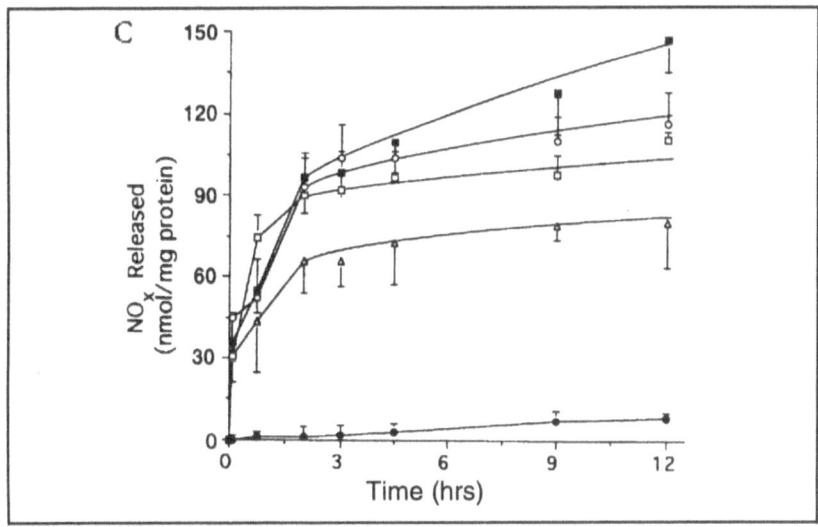

Fig 2.5C. Effect of shear stress on cumulative (NO_2^- and NO_3^{2-}) NO_x release. Cultures were exposed to no flow (●), flow with 1.8 dyn/cm² shear stress ( ), 6 dyn/cm² ( ), 12 dyn/cm² ( ), or 25 dyn/cm² (■). Measurements were taken on three separate cultures. Each value represents mean +/- SE. At all time points and shear stresses examined exposure to flow significantly elevated NO_x. cGMP was significantly elevated by all levels of shear after 5 and 30 min. However, cGMP was elevated only by 12 and 25 dyn/cm₂ beyond 3 h ($p < 0.05$, ANOVA). With permission from Kuchan MJ and Frangos JA. Role of calcium and calmodulin in flow-induced nitric oxide production in endothelial cells. Am J Physiol 1994; 266:C628-36.

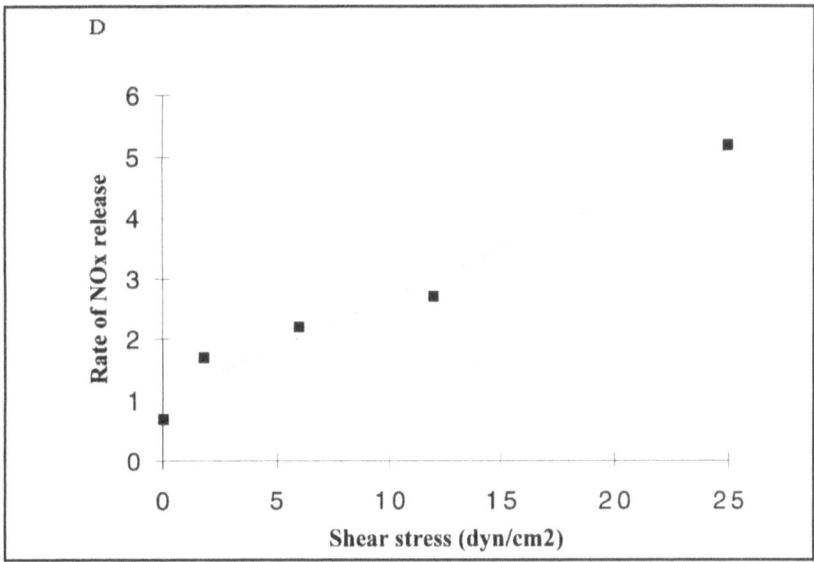

Fig. 2.5D. Effect of shear stress on the average NO_x release rate 2-12 h after the initiation of flow. Rate of release expressed as nmol/mg protein/h.

A

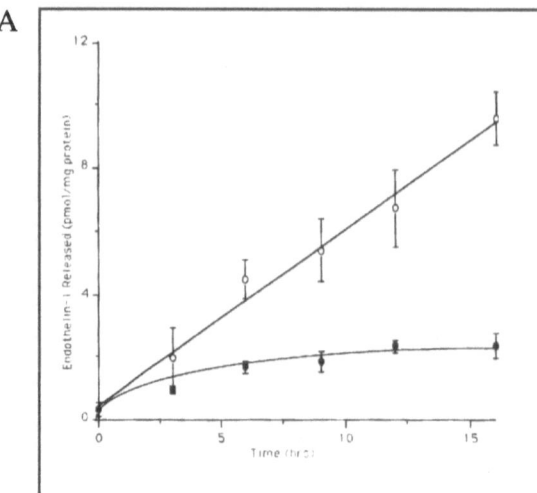

B

C

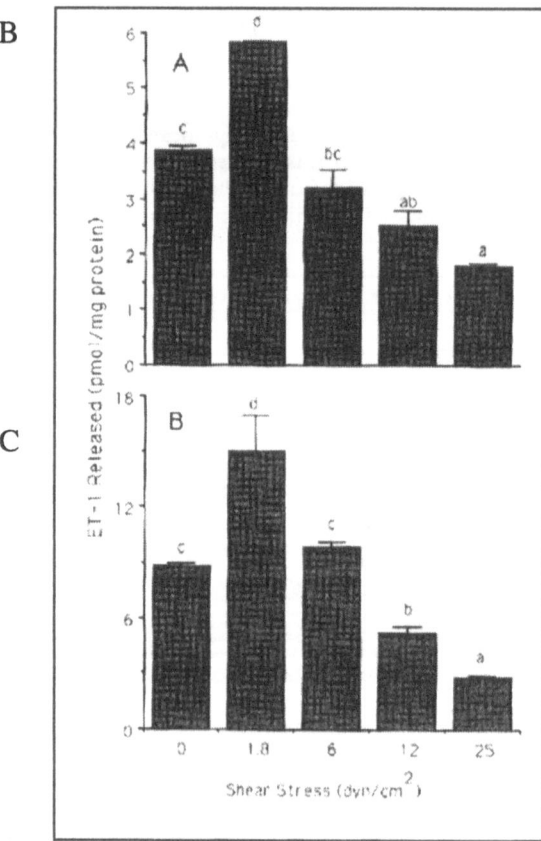

Fig. 2.6. ET-1 production rate by HUVEC varies as a function of time of exposure (Panel A) and magnitude of shear stress (Panel B and C). Panel A shows cumulative ET-1 release by HUVEC under stationary (O) and steady laminar flow conditions of 10 dyn/cm^2 (●). The dose response relationship between shear stress and release of ET-1 after 6 h (Panel B) and 16 h (Panel C). Kuchan MJ and Frangos JA. Shear stress regulates ET-1 release via protein kinase C and cGMP in cultured endothelial cells. With permission from Kuchan MJ, Frangos JA. Am J Physiol 1993; 264:H150-6.

NO

Like PGI_2 release, NO release is increased by the application of laminar fluid flow to cultured endothelial cells (Fig. 2.5, Panel C and D). Although PGI_2 and NO are members of two distinct classes of compounds synthesized by different enzymes, the kinetics of their release by HUVEC are strikingly similar (compare Panels A and B to C and D of Fig. 2.5). With the onset of flow, there is an initial burst in production lasting approximately 10-30 min followed by a nearly constant production rate which, while not as great as the initial burst, is several times greater than basal production. This similarity in production kinetics suggests that either the same factors regulate the production of each compound, or that one compound regulates the release of the other. A recent study supports the second hypothesis. In this study, Davidge et al observed that NO stimulated the activity, but not the synthesis, of prostaglandin H synthase.[97]

Control of smooth muscle proliferation

The supernatant of endothelial cells exposed to laminar flow inhibits mesangial cell proliferation through an unknown mechanism.[98] It is not known whether this supernatant also modifies the proliferation of smooth muscle cells. One drawback of a system in which the supernatant collected from flow-exposed endothelial cells is used to test for the presence of compounds with mitogenic or antimitogenic properties is that unstable compounds would likely be overlooked. To identify unstable mediators of proliferation, the smooth muscle cells must constantly be exposed to medium that has just been in contact with endothelial cells exposed to flow.

Little work has been conducted on this topic in vitro, presumably because of the difficulties of culturing endothelial cells and smooth muscle cells in close proximity while exposing the endothelial cells, but not the smooth muscle cells, to fluid flow. It is important that the smooth muscle cells not be exposed directly to flow, as flow is known to inhibit their proliferation.[99] A co-culture system developed by Nerem and coworkers may serve as a model for studying this problem. In this system, a layer of endothelial cells are grown on a collagen gel containing smooth muscle cells. In this system fluid flow exhibited a slight inhibition of endothelial and smooth muscle cell proliferation.[100]

Control of transvessel transport

Laminar fluid flow has been shown to affect the transendothelial permeability of both albumin and water across a confluent monolayer of bovine aorta endothelial cells grown on polycarbonate filters. Fluorescently labeled albumin added to the luminal side of the filters was allowed to diffuse across the endothelial monolayer in the absence of a hydrostatic pressure drop across the filter, which would drive convection. The initiation of steady laminar flow with an average shear stress of 10 dyn/cm² resulted in an approximately 10-fold increase in albumin permeability after 30 min. The elevated permeability persisted until flow was terminated after a total of 60 min. Permeability returned to basal levels within 30 min after flow cessation. The initiation of steady flow with an average shear stress of 1 dyn/cm² resulted in a smaller increase in permeability (approximately sixfold), requiring 1 hour to reach its maximum and another hour to return to its basal level after cessation of flow.

In a similar experimental system with a confluent monolayer of BAEC on a polycarbonate filter, the initiation of steady laminar flow resulted in a dose-dependent increase in hydraulic conductivity (Fig.2.7, Panels A-E). Shear stresses as low 0.5 dyn/cm² resulted in an increase in hydraulic conductivity compared to stationary controls. Fluid flow with a wall shear stress of 20 dyn/cm² elicited a threefold increase in hydraulic conductivity after a three-hour exposure to flow.[26]

Different flow conditions result in different responses

Flow conditions are poorly defined in many studies that attempt to describe the biological responses of endothelial cells to fluid flow. This is unfortunate, as the magnitude, duration and rate of change of flow rate, as well as the frequency and amplitude of flow pulsatility, may play an important role in determining the magnitude and type of response. The role of each of these characteristics of the fluid flow in regulating cellular responses will be discussed in more detail below.

Magnitude

Several reports have indicated that the magnitude of steady laminar flow dictates endothelial cell response. The steady state production rate of PGI₂ by HUVEC increased linearly with shear stress from zero to 24 dyn/cm² (Fig. 2.5, Panel B).[101] The production rate of t-PA by HUVEC exposed to steady laminar flow also increased with increasing shear stress of a similar range (4-25 dyn/cm²).[54] More recently, the production rates of NO[70] (Fig. 2.5, Panel B) and of mRNA for nitric oxide synthase (NOS)[102] have been shown to increase with increasing shear stress.

In contrast to the above compounds, whose production rate increases monotonically with shear stress (at least in studies representing physiological shear stress), the ET-1 response to shear stress is biphasic. Before 1992, the effect of fluid flow on ET-1 production by endothelial cells was controversial, with some laboratories reporting flow-induced increases[103,104] while others reported decreases.[105] These controversies were resolved by a systematic study revealing that ET-1 release is dependent upon both magnitude and duration of flow.[96] For example, cumulative release of ET-1 over 6 hours increases with increasing shear stress until reaching a maximum 50% increase at 1.8 dyn/cm². Larger levels of shear stress result in a progressive decline in cumulative ET-1 release to values 50% below basal levels (Fig. 2.6). These data emphasize the importance of the magnitude of shear stress in determining the biological response, especially in responses observed after hours of stimulation by flow.

Duration

A dramatic example illustrating the importance of the time of exposure to fluid flow as a determinant of the biochemical response is Hseih's work on flow-induced gene expression. The initiation of steady laminar flow with a wall shear stress of 16 dyn/cm² results in a 24-fold increase in the amount of c-fos mRNA in HUVEC after 30 min. Remarkably, *c-fos* mRNA levels return to basal levels after 30 additional minutes, even in the presence of continual steady laminar flow.

Kuchan et al showed that the flow-induced stimulation of NO production can be conceptualized as having two phases: an acute phase lasting for about 45 min after the initiation of flow, and a chronic phase beginning about 1 hour after flow initiation and lasting at least 48 hours. In the acute phase, NO production rate was about 10 times higher than control levels, whereas chronic NO production was about four times higher than control levels, again demonstrating that the duration of the applied force regulates the response. Even though flow resulted in both acute and chronic increases in NO production, the biochemical mechanisms mediating these responses are very different. The acute response is Ca²⁺-dependent and G-protein dependent, whereas the chronic response is independent of both factors. Taken together, these data emphasize that the duration of stimulus is an important determinant of the response as well as the mechanisms regulating the response.[96,106]

Rate of change

In addition to the magnitude and duration of shear stress, the rate of change of shear stress may be equally important or, in some cases, more important. Rapid transition from stationary to flow conditions results in a burst of NO production by HUVEC lasting about 30 min as measured by extracellular accumulation of NOx (the sum of nitrate and nitrite, the stable oxidation products of NO). The burst in NO is largely insensitive to the magnitude of shear stress, with a step change in flow rate from 0 to 1.8 or 25 dyn/cm² producing approximately the same response. The early response of endothelial cells to the initiation of steady flow is mimicked by a three second expo-

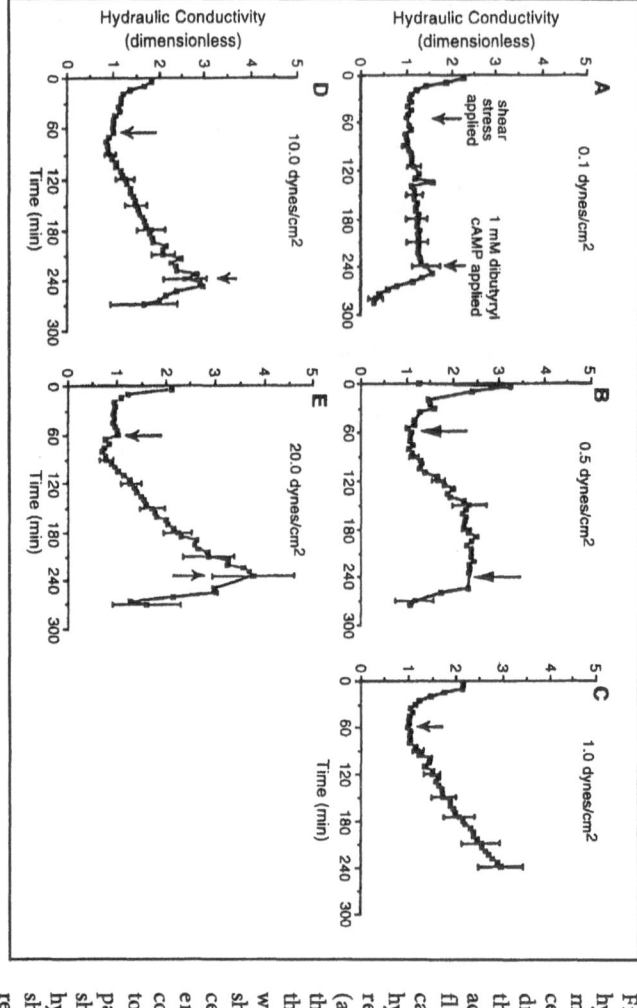

Fig. 2-7. Laminar fluid flow increases hydraulic conductivity across a confluent monolayer of bovine aortic endothelial cells. At time 0, a hydrostatic pressure drop of 10 cm of water was applied across the monolayer, resulting in fluid flow across the monolayer. By measuring this flow, the hydraulic conductivity was calculated. Over the first half hour, the hydraulic conductivity decreased until it reached a steady state value (approximately 50% that of the initial) that served as a base line against which the results from individual experiments were normalized. After a total of one hour, shear stress was applied to the endothelial cells by rotating a disc parallel to the endothelial monolayer. For a more complete description of the device used to apply shear stress, see the section on parallel-disc viscometers in chapter 1. A shear stress of 0.1 dyn/cm² did not affect hydraulic conductivity (Panel A). Larger shear stresses from 0.5 to 20.0 dyn/cm² resulted in a dose-dependent increase in hydraulic conductivity (Panels B-E). The addition of 1 mM dibutyryl cAMP 3 hours after initiating flow conditions decreased hydraulic conductivity (Panels A, B, D, and E), indicating that the flow-induced increases in conductivity are reversible. From Sill HW, Chang YS, Artman JR et al. Shear stress increases hydraulic conductivity of cultured endothelial monolayers. Am J Physiol 1995; 268: H535-43.

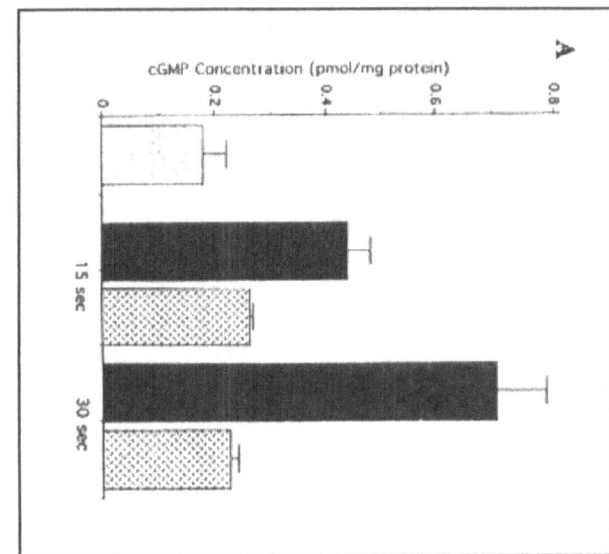

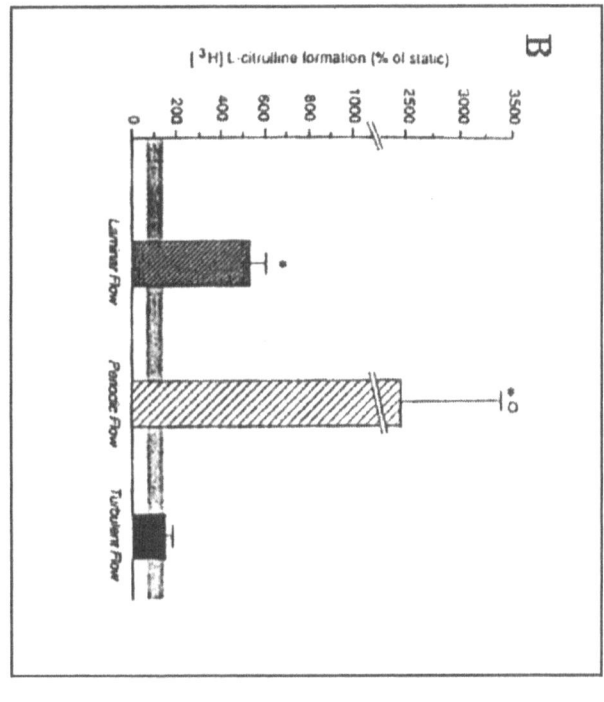

Fig. 2.8. The effect of varying the rate of change of shear stress on NO production by endothelial cells. Panel A shows the effect of step (dark bars) or ramp (stippled bars) changes in flow from 0.0 to 1.8 dyn/cm² on intracellular cGMP, and index of NO concentration, compared to stationary controls (light bar). From Kuchan MJ, Frangos JA. Role of calcium and calmodulin in flow-induced nitric oxide production in endothelial cells. Am J Physiol 1994; 266:C628-36. Both periodic flow (square wave from 2 to 8 dyn/cm² with 15 min cycles) and steady laminar shear stress of 8 dyn/cm² increased cumulative tritiated L-citrulline production, an index of NO production, over 6 hours compared to stationary controls (stippled horizontal bar). Turbulent flow with an average shear stress of 8 dyn/cm² did not increase NO production (Panel B). From Noris M, Morigi M, Donadelli R et al. Nitric oxide synthesis by cultured endothelial cells is modulated by flow conditions. With permission from Kuchan MJ, Frangos JA, Am J Physiol 1993; 264:H150-6; and Noris M et al, Circ Res 1995; 76(4):536-43.

sure to flow followed by a return to stationary conditions (John Frangos, personal communication). Interestingly, if instead of a step change, flow is gradually increased to its maximum value over 15 sec, the burst of NO production is largely inhibited; an increase over 30 sec results in complete inhibition (Fig. 2.8, Panel A).[70] Taken together, these data indicate that the rate of change in flow is the important parameter in determining the acute (on the order of minutes) NO response, while the chronic response is dependent on the magnitude of flow. The physiological significance of endothelial cells being able to respond to both the rate of change and the magnitude of shear stress on different time scales suggests that the vascular system should be able to respond quickly to small but abrupt changes in perfusion as well as chronic changes in flow.

As might be predicted by the above data, subsequent studies have revealed that flow changing abruptly from 2 to 8 dyn/cm^2 in square wave with a 15-min cycle, which can be conceptualized as periodic step changes superimposed on a steady flow, is more stimulatory than steady flow after 6 hours (Fig 2.8, Panel B).[102] Surprisingly, exposure to turbulent flow with an average shear stress of 8 dyn/cm^2, which would be expected to exert a shear stress varying rapidly as a function of time, did not stimulate NO production, suggesting that extremely rapid variations in flow are less stimulatory than more slowly varying flow.[102] This hypothesis is consistent with the observations of Hutcheson et al that NO production varies as a function of the frequency (0.1-12 Hz) of flow, with peak response occurring at approximately 5 Hz.[107]

While many different biological responses apppear to be sensitive to the magnitude, duration, and rate of change of the applied physical force, NO production by endothelial cells in response to fluid flow clearly illustrates this principle. Figure 2.9 summarizes results from the work of Kuchan et al,[70] Noris et al,[102] and Hutchenson et al on the effect of fluid flow on NO release by endothelial cells.[107]

Endothelial Cell Responses to Pressure

In vivo, pressure exerts a force orders of magnitude larger than that of shear stress. Elevated blood pressure is a commonly accepted risk factor for cardiovascular diseases. Despite the potential importance in human health, the effects of pressure on cells in the vascular system have only recently been investigated.

Bovine pulmonary artery endothelial cells were exposed to a sustained hydrostatic pressure generated by submerging the cells in 10 cm of medium. The increased hydrostatic pressure stimulated proliferation and elongation of the cells compared to control cultures submerged in 3 cm of medium, apparently secondary to bFGF release by the endothelial cells.[108] The increases

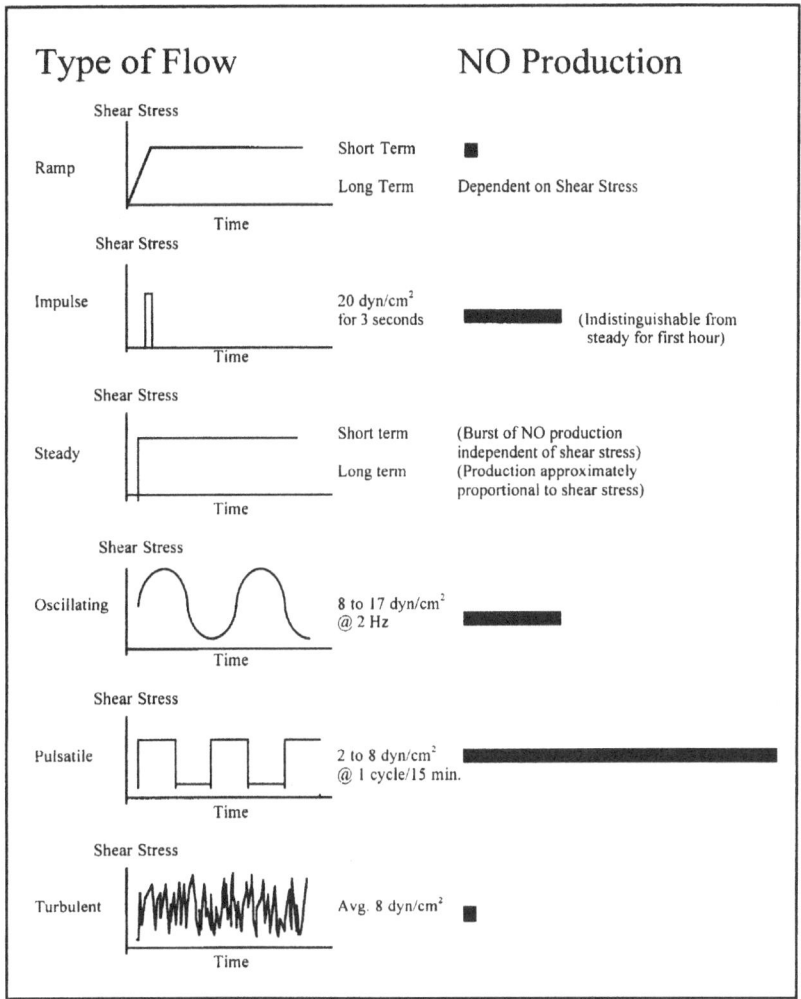

Fig. 2.9. A summary of the results from three papers on the effects of different flow patterns on NO production by cultured endothelial cells. The graphs on the left illustrate each flow pattern while corresponding NO production is described on the right. The length of the filled horizontal bars is roughly proportional to measured NO production. As the three papers used different indices of NO production, it is difficult to precisely compare NO production. See references in Figure 2.8 for more details.

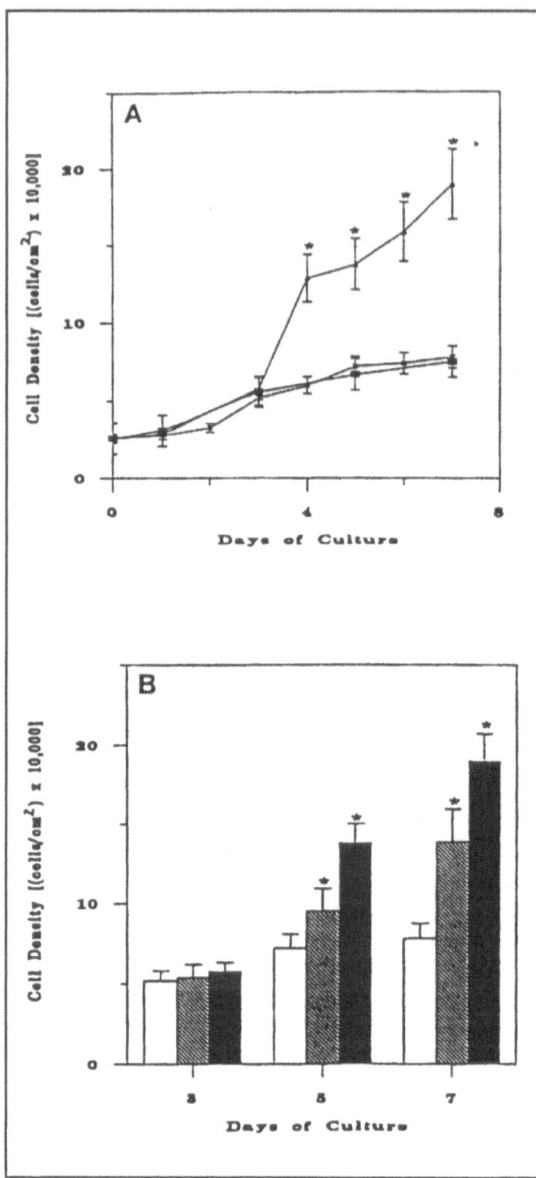

Fig. 2.10. Time course of bovine pulmonary artery endothelial cell proliferation. Panel A depicts growth curves of endothelial cells cultured in DMEM supplemented with 20% FBS under the following conditions: control (pressure = 3 mm H_2O) with 2 ml of medium without replacement for the duration of the experiment (x); control (pressure = 3 mm H_2O) with medium replacement equal to the volume of a 10 cm H_2O hydrostatic head (■); and 10 cm H_2O pressure conditions (★). The bar graph (Panel B) shows endothelial cell density after exposure to pressures of 3 mm H_2O (open bars), 1.5 cm H_2O (striped bars), and 15 cm H_2O (filled bars) after 3, 5, and 7 days. Values are mean + standard deviation; n = 3. Significant differences from control (P < 0.05) are indicated by asterisks. From Acevedo et al. Morphological and proliferative response of endothelial cells to hydrostatic pressure: role of fibroblast growth factor. J Cell Physiol 1993; 157:603-14. Copyright © 1993. Reprinted by permission of Wiley-Liss, Inc., a subsidiary of John Wiley & Sons, Inc.

in proliferation were observable after 4 days and continued for at least 1 week (Fig. 2.10). Several experiments were conducted in an attempt to exclude artifacts resulting from differences in nutrient availability, pH, pO_2 and pCO_2 between cells cultured under 3 cm and 10 cm of medium. The physiological significance of these responses to a small change in pressure (about 8 mm of Hg) is not obvious. In another study, larger pressure increases resulting from elevated atmospheric pressure resulted in slight (less than 20%) increases in ^{3}H-thymidine incorporation, an indicator of DNA synthesis, at 50 and 100 mm Hg after 1 day. A physiologically high pressure of 200 mm of Hg resulted in an approximate doubling in ^{3}H-thymidine incorporation.[109]

Increasing atmospheric pressure by the addition of helium results in a dose-dependent increase in the production of the vasoconstrictor ET-1, with a maximal observed increase of approximately 20% at 160 mm Hg above normal atmospheric pressure and a half-maximal response occurring between 40 and 80 mm Hg.[110] This pressure-induced response was abrogated by inhibitors of phospholipase C and protein kinase C but not by Gd^{3+}, an inhibitor of some classes of stretch-activated ion channels. In addition, a pressure of 80 mm Hg decreased NO production by approximately 20%. The simultaneous increase in production of ET-1, a vasoconstrictor, and decrease in production of NO, a vasodilator, suggests that pressure stimulates a coordinated response by the endothelial cells that may result in pressure-induced vasoconstriction in vivo. See chapter 6 for further discussion of pressure-induced vasoconstriction (myogenic response). Though the changes in ET-1 and NO release due to pressure are statistically significant, they are much smaller than the changes observed with variations in fluid flow,[96] suggesting that pressure may not be the major determinant of ET-1 and NO release in vivo.

Endothelial Cell Responses to Cyclic Strain

The effects of cyclic strain or cyclic stretch on endothelial cells have been recently reviewed.[31,111,112] In addition to shear stress and pressure, endothelium is exposed to forces resulting from cyclic strain driven by oscillating blood pressure. The effects of cyclic strain often are studied in vitro by culturing cells on one side of a flexible surface that can be deformed by the application of pressure or vacuum to the other side. Such in vitro studies have demonstrated that increases or decreases in the frequency of cyclic strain activate phospholipase C (PLC), resulting in increases in the concentration of IP_3 and DAG in bovine aorta endothelial cells.[73,113] The initiation of cyclic strain, however, does not universally stimulate second-messenger pathways, as evidenced by its inability to modify PAI-1[114] production or cAMP concentration[115] in human saphenous vein endothelial cells.

Cyclic strain,[116] like shear stress,[75] causes endothelial cells to elongate and align by a mechanism dependent on the actin filament system. A significant difference between the responses is that shear stress results in alignment parallel to the applied force while cyclic strain causes alignment perpendicular to force. In vivo these forces often act perpendicularly, suggesting that cyclic strain and shear stress may work synergistically to align endothelial cells parallel to blood flow.

Other examples exist of shear stress and cyclic strain potentially working together to stimulate endothelial cells to produce the same response. Both shear stress and cyclic strain stimulate cultured endothelial cells to increase their production of PGI_2,[52,117] t-PA,[54,56,114] and $NO^{32,70}$ (Fig. 2.11, Panels A and B). It is not known if these physical stimuli have an additive effect or if either is adequate for maximal physical force-induced stimulation. A recent report again highlights the similarities between flow- and cyclic strain-induced stimulation of endothelial cells. Mechanical strain of 1 Hz induces monocyte chemotactic protein-1 (MCP-1) gene expression in endothelial cells (Fig. 2.12, Panel A).[118] This increase in MCP-1 mRNA is translated into increased protein concentration in conditioned medium, which is further evidenced by increased chemotactic activity in vitro (Fig. 2.12, Panel B).[118] As noted previously, MCP-1 mRNA is transiently increased by steady laminar fluid flow.

The striking similarities in the biological responses elicited by fluid flow and cyclic strain suggest that the way endothelial cells perceive these two physical forces may be similar. That is, the actual mechanical signal stimulating

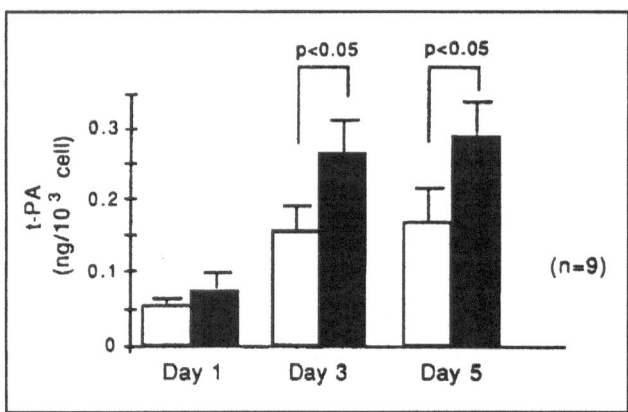

Fig 2.11A. Human saphenous vein endothelial cells subjected to cyclic strain (24% maximum strain at 1 Hz) increased their production of t-PA (filled bars) compared to stationary cells (open bars) (Panel A). Data are shown as mean + standard deviation ($n \geq 6$). With permission from Iba et al. J Surg Res 1991; 50: 457-60.

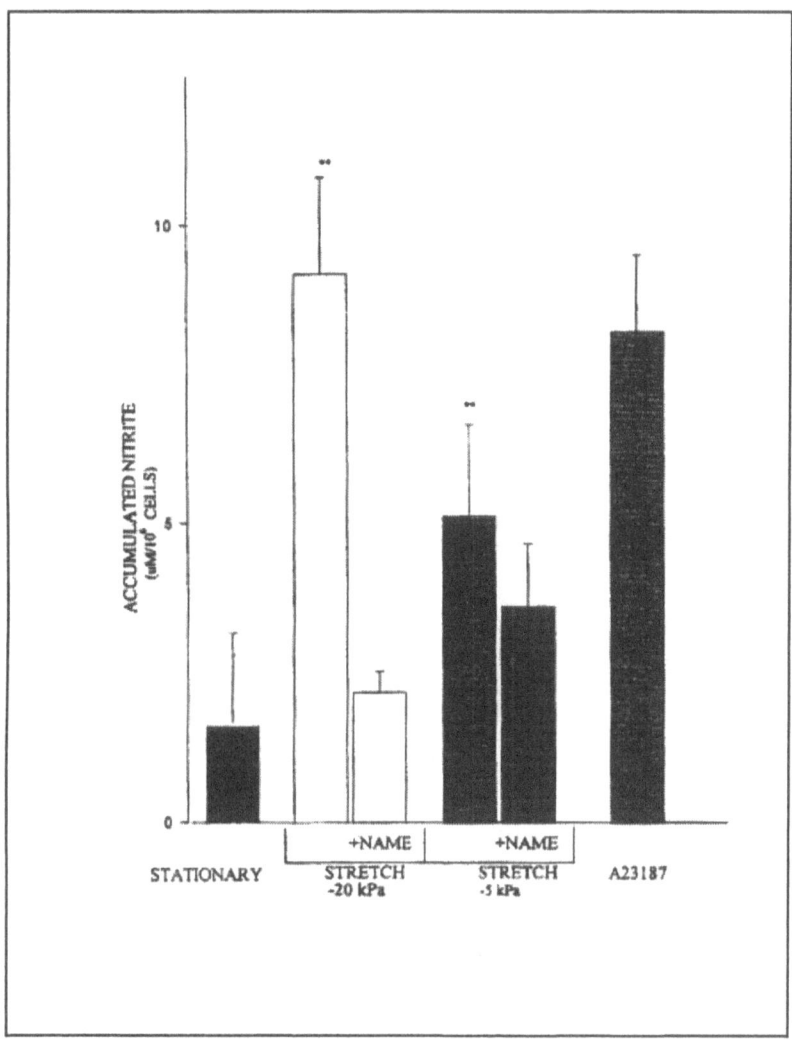

Fig. 2.11B. Panel B depicts NO production by bovine aortic endothelial cells subjected to cyclic strain (24% maximum strain at 1 Hz generated by applying -20 kPa or 10% maximum strain at 1 Hz generated by applying -5 kPa) or stationary conditions for 24 hrs. NO production was inferred by measuring accumulated nitrate in the conditioned medium using the Griess reaction. Strain-induced nitrite accumulation was abrogated by the nitric oxide synthase inhibitor L-NAME. Strain-induced nitrite accumulation was comparable to that generated by the endothelial cells when exposed to the calcium ionophore A23187, a positive control. Data are shown as mean + standard deviation (n ≥ 6). With permission from Awolesi MA. Cyclic strain increases endothelial nitric oxide synthase activity. Surgery 1994; 116(2):439-45.

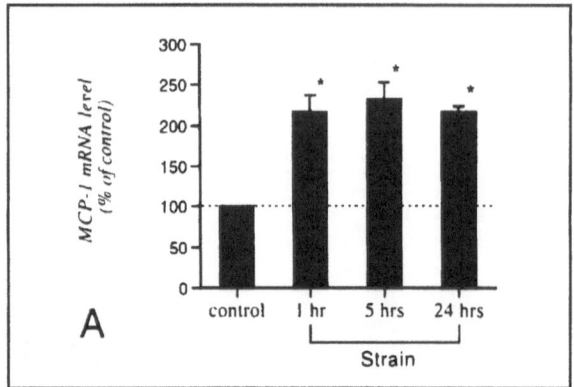

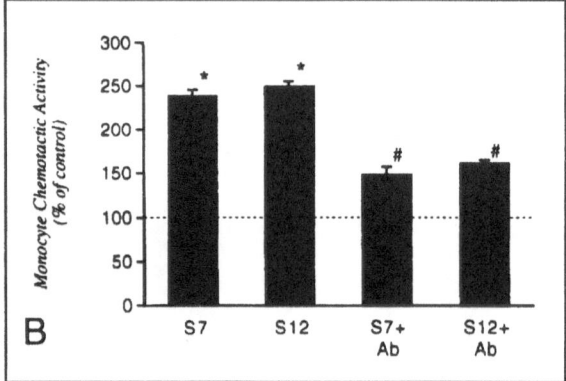

Fig 2.12. Cyclic strain (1 Hz) increases the MCP-1 gene expression in HUVEC as revealed by densitometric analysis of Northern blots from three separate experiments (Panel A, top). After discontinuing mechanical stimulation, MCP-1 mRNA levels return to control values by 3 hrs (data not shown). The increase in mRNA corresponds to increased monocyte chemotactic activity (Panel B, bottom). Conditioned medium from endothelial cells exposed to cyclic strain for 7 or 12 hours possessed increased monocyte chemotactic activity in Boyden chamber experiments compared to controls. This increase in activity was substantially reduced by preincubating conditioned medium with antibodies to MCP-1. With permission from Wang et al. Circ Res 1995; 77:294-302.

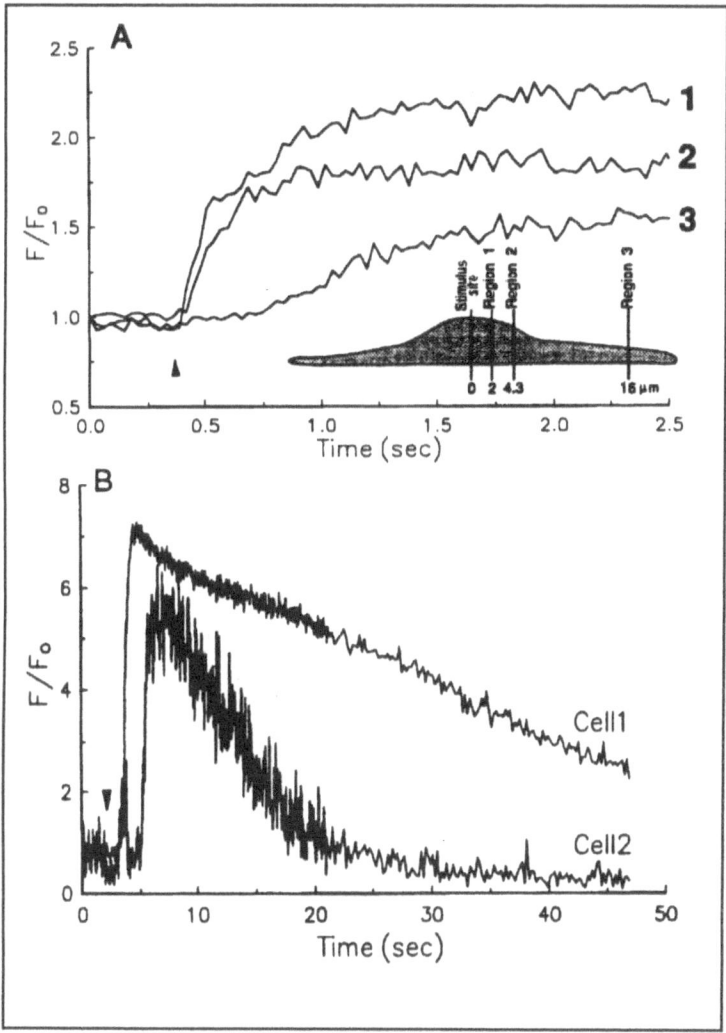

Fig 2.13. Tracings obtained from mechanically activated endothelial cells demonstrating temporal and spatial gradients of [Ca²⁺]ᵢ. A: average fluorescence intensities normalized to initial fluorescence intensity (F/Fo) within a nuclear sample region (1), a perinuclear sample region (2), and a peripheral cytoplasmic region (3) (as indicated schematically in inset) are shown for an endothelial cell responding to mechanical activation. Region 1 is first to achieve maximal Ca²⁺ levels, whereas regions 2 and 3 achieve their maximal response with a time delay relative to region 1 of 150 and 600 ms, respectively. A, inset: schematic of endothelial cell showing radial distances of regions 1, 2 and 3 from stimulus site. B: mechanical stimulation of cell 1 resulted in increased fluo-3 fluorescence, which was followed by an increase in an unstimulated neighbor (cell 2). Time lag of ~1.5 s was required for cellular communication between cells, with the mechanically activated cell having already achieved maximal response. With permission from Sigurdson WJ et al. Am J Physiol 1993; 264:H1745-52.

the cells may be the same. The common mechanical denominator between fluid flow and cyclic strain, however, is not evident. One hypothesis is that fluid flow stretches the cell, resulting in the same biological responses as cyclic strain. This hypothesis, however, is inconsistent with the experimental observation that fluid flow results in imperceptibly small deformations of the cell while typical strains used in stretching experiments are approximately 10%.

Endothelial Cell Responses to Mechanical Perturbation

Direct mechanical deformation of regions of the plasma membrane of single endothelial cells produced a transient increase in intracellular calcium which returned to basal levels within 30 sec, as measured by fluorescence microscopy (Fig. 2.13). This increase in intracellular calcium was not inhibited by lowering the calcium concentration of the bathing medium, suggesting that the increase in calcium results from a release of intracellular stores.[119]

Summary

Due to the involvement of endothelial cells in vascular disease and their constant exposure to varied mechanical forces in vivo, the endothelial cell perhaps is the cell type whose responses to physical forces have been studied most extensively. The important physiological functions performed by endothelial cells include maintaining an antithrombogenic environment while controlling vasoregulation, smooth muscle cell proliferation, and transvessel transport.

Cardiac and arterial endothelial cells are exposed to cyclic strain that accompanies the cardiac cycle, while venous endothelial cells are strained by venous pumping. Fluid flow, probably acting through shear stress, regulates many physiologically significant endothelial functions, including biochemical second messenger systems and gene expression (Table 2.1), in addition to the physiological responses discussed above.

The magnitude, duration and rate of change of flow rate, as well as the frequency and amplitude of pulsatility of the flow, may play important roles in determining the magnitude and type of cellular response. Pressure ranging from <1 to 200 mm Hg have been reported to modify endothelial cell activity in vitro, and direct mechanical perturbation also results in cellular responses. Cyclic strain elicits many of the same biological resonses as fluid flow, suggesting that the mechanisms by which endothelial cells perceive these two forces may be similar. However, the common mechanical denominator between fluid flow and cyclic strain is not evident.

References

1. Glagov S et al. Hemodynamic and atherosclerosis: Insights and perspectives gained from studies of human arteries. Arch Path Lab Med 1988; 112:1018-31.
2. Glagov S, Ozoa AK. Significance of the relatively low incidence of atherosclerosis in the pulmonary, renal, and mesenteric arteries. NY Acad Sci 1968; 149:940-55.
3. Boeynaems JM, Pirotton S. Regulation of the vascular endothelium: signals and transduction mechanisms. Austin: RGLandes, 1994.
4. Marcum JA et al. Cloned bovine aortic endothelial cells synthesize anticoagulantyl activite heparan sulfate proteoglycan. J Biol Chem 1986; 261:7507-17.
5. Levin EG, Loskutoff DJ. Cultured bovine endothelial cells produce both urokinase and tissue-type plasminogen activators. J Cell Biol 1982; 94:631-6.
6. Vane JR, Gryglewski RJ, Botting RM. The endothelium as a metabolic and endocrine organ. TIPS 1987; 8:491-496.
7. Palmer RMJ, Ferrige AG, Moncada S. Nitric oxide release accounts for the biological activity of endothelium-derived relaxing factor. Nature 1987; 327:524-6.
8. Ignarro LJ et al. Endothelium-derived relaxing factor produced and released from artery and vein is nitric oxide. Proc Natl Acad Sci USA 1987 84: 9265-9.
9. Moncada S, Vane JR. Pharmacology and exogenous roles of prostaglandin endoperoxides, thromboxane A2 and prostacyclin. Pharmacol Rev 1978; 30:293-331.
10. Radomski MW, Palmer RMJ, Moncada S. Endogenous nitric oxide inhibits human platelet adhesion to vascular endothelium. Lancet 1987; ii:1057-8.
11. Radomski MW, Palmer RMJ, Moncada S. The anti-aggregating properties of vascular endothelium: interactions between prostacyclin and nitric oxide. Br J Pharmacol 1987; 92:639-46.
12. Murad F. Cyclic guanosine monophosphate as a mediator of vasodilation. J Clin Invest 1986; 78:1-5.
13. Bolotinal VS et al. Nitric oxide directly activates calcium dependent potassium channels in vascular smooth muscle. Nature 1994; 368:850-3.
14. Gajdusek C et al. An endothelial cell-derived growth factor. J Cell Biol 1980; 85:467-72.
15. DiCorleto PE, Bowen-Pope DF. Cultured endothelial cells produce a platelet-derived growth factor-like protein. Proc Natl Acad Sci USA 1983; 80:1919-23.
16. Komoro I et al. Endothelin stimulates c-fos and c-myc expression and proliferation of vascular smooth muscle cells. FEBS Lett 1988; 238:249-52.

17. Nakaki T et al. Endothelin-mediated stimulation of DNA synthesis in vascular smooth muscle cells. Biochem Biophys Res Commun 1989; 158:880-3.
18. Castellot JJ et al. Cultured endothelial cells produce a heparin-like inhibitor of smooth muscle cell growth. J Cell Biol 1981; 90:372-9.
19. Garg UC, Hassidand A. Nitric oxide generating vasodilators and 8-bromo-cyclic GMP inhibit mitogenesis and proliferation of cultures rat vascular smooth muscle cells. J Clin Invest 1990; 83:1018-31.
20. Sarkar R, Webb RC, Stanley JC. Nitric oxide inhibition of endothelial cell mitogenesis and proliferation. Surgery 1995; 118(2):274-9.
21. Gooch KJ, Dangler CA, Frangos JA. The role of endogenous, exogenous, and flow-induced nitric oxide in regulating endothelial cell proliferation. J Cell Physiol, Submitted.
22. Lombardi T et al. Endothelial diaphragmed fenestrae: In vitro modulation by phorbol myristate acetate. J Cell Biol 1986; 102:1965-70.
23. Milici AJ et al. Transcytosis of albumin in capillary endothelium. J Cell Biol 1987; 105:2603-12.
24. Predescu D, Palade GE. Plasmalemmal vesicles represent the large pore system of continuous microvascular endothelium. Am J Physiol 1993; 265:H725-33.
25. Majno G, Palade GE. Studies on inflammation. The effect of histamine and serotonin on vascular permeability: An electron microscopic study. Biochem Cytol 1961; 11:571-85.
26. Sill HW et al. Shear stress increases hydraulic conductivity of cultured endothelial monolayers. Am Journal Physiol 1995; 268:H535-43.
27. Grega GJ. Contractile elements in endothelial cells as potential target for drug action. TIPS 1986; 7:452-7.
28. Wysolmerski R, Langoff D. Involvement of myosin light-chain kinase in endothelial cell retraction. Proc Natl Acad Sci USA 1990; 87:16-20.
29. Davies PF. Flow-mediated endothelial mechanotransduction. Physiol Rev 1995; 75(3): 519-60.
30. Berthiaume F, Frangos JA. Effect of flow on anchorage-dependent mammalian cells-secreted products. In: Frangos JA, Ed. Physical Forces and the Mammalian Cell. Academic Press: San Diego, 1993:139-192.
31. Patrick CW, McIntire LV. Shear stress and cyclic strain modulation of gene expression in vascular endothelial cells. Blood Purif 1995; 13(3-4):112-24.
32. Cooke JP et al. Flow activates an endothelial potassium channel to release an endogenous nitrovasodilator. J Clin Invest 1991; 88:1663-71.
33. Jacobs E et al. Shear-activated channels in cell-attached patches of aortic endothelial cells. FASEB J, 1993.
34. Oleson S, Clapham D, Davies P. Hemodynamic shear stress activates a K^+ current in vascular endothelial cells. Nature 1988; 331:168-170.
35. Nakache M, Gaub H. Hydrodynamic hyperpolarization of endothelial cells. Proc Natl Acad Sci USA 1988; 85:1841-3.

36. Alevriadou B et al. Effect of shear stress on $^{86}Rb^+$ efflux from calf pulmonary artery endothelial cells. Ann Biomed Engineer 1993; 21:1-7.

37. Lansman J, Hallam T, Rink T. Single stretch-activated ion channels in vascular endothelial cells as mechanotransducers? Nature 1987; 325:811-2.

38. Goligorsky M. Mechanical stimulation induces Ca^{++}_i transients and membrane depolarization in cultured endothelial cells. FEBS Lett 1988; 240:59-64.

39. Ziegelstein R, Cheng L, Capogrossi M. Flow dependent cytosolic acidification of vascular endothelial cells. Science 1992; 258:656-9.

40. Ando J, Komatsuda T, Kamiya A. Cytoplasmic calcium responses to fluid shear stress in cultured vascular endothelial cells. In Vitro Cell Dev Biol 1988; 24:871-7.

41. Geiger R et al. Flow-induced calcium transients in single endothelial cells: spatial and temporal analysis. Am J Physiol 1992; 262:C1411-7.

42. Shen J et al. Fluid shear stress modulates cytosolic free calcium in vascular endothelial cells. Am J Physiol 1992; 262:C384-90.

43. Schwarz G et al. Shear stress-induced calcium transients in endothelial cells from human umbilical cord veins. J Physiol 1992; 458:527-38.

44. Frangos J, Kuchan M. Fluid flow activates G proteins that are coupled to Ca-dependent and -independent EDRF production in cultured endothelial cells (abstract). FASEB J 1991; 5:A1820.

45. Taylor W et al. Characterization of the release of endothelial-derived nitrogen oxides by shear stress (abstract). FASEB J 1991; 5:A1727.

46. Nishida K et al. Molecular cloning and characterization of the constitutive bovine aortic endothelial cell nitric oxide synthase. J Clin Invest 1992; 90:2092-96.

47. Bodin P, Bailey D, Burnstock G. Increased flow-induced ATP release from isolated vascular endothelial cells but not smooth muscle cells. Br J Pharmacol 1991; 103(1):1203-5.

48. Bodin P et al. Chronic hypoxia changes the ratio of endothelin to ATP release from rat aortic endothelial cells exposed to high flow. Proc R Soc London B 1992; 247:131-5.

49. Milner P et al. Endothelial cells cultured from human umbilical vein release ATP, substance P and acetylcholine in response to increased flow. Proc R Soc London B 1990; 241:245-8.

50. Ralevic V et al. Substance P is released from the endothelium of normal and capsaicin-treated hind limb vasculature in vivo by increased flow. Circ Res 1990; 66:1178-83.

51. Berthiaume F, Frangos J. Flow-induced prostacyclin production is mediated by a pertussis toxin-sensitive G protein. FEBS Lett 1992; 308(3):277-9.

52. Frangos JA et al. Flow effects on prostacyclin production by cultured human endothelial cells. Science 1985; 227:1477-9.

53. Grabowski EF, Jaffe EA, Weksler BB. Prostacyclin production by cultured endothelial cell monolayers exposed to step increases in shear stress. J Lab Clin Med 1985; 103:36-43.

54. Diamond SL, Eskin SG, McIntire LV. Fluid flow stimulates tissue plasminogen activator secretion by culture human endothelial cells. Science 1989; 243:1483-5.

55. Skarlatos S, Hollisand T. Cultured bovine aortic endothelial cells show increased histamine metabolism when exposed to oscillatory shear stress. Atherosclerosis 1987; 64:55-61.

56. Diamond SL et al. Tissue plasminogen activator mRNA levels increase in cultures human endothelial cells exposed to laminar shear stress. J Cell Physiol 1990; 143:364-71.

57. Lan Q, Mercurius K, Davies P. Stimulation of transcription factors NFkB and AP-1 in endothelial cells subjected to shear stress. Biochem Biophys Res Commun 1994; 201:950-6.

58. Hsieh H, Li N, Frangos J. Pulsatile and steady flows increase proto-oncogenes c-fos and c-myc mRNA levels in human endothelial cells (abstract). FASEB J 1991; 5:A1820.

59. Ohtsuka A et al. The effect of flow on the expression of vascular adhesion molecule-1 by cultured mouse endothelial cells. Biochem Biophys Res Commun 1993; 193:303-10.

60. Hsieh H, Li N, Frangos J. Shear-induced platelet derived growth factor gene expression in human endothelial cells is mediated by protein kinase C. J Cell Physiol 1992; 150:552-8.

61. Mitsumata M et al. Fluid shear stress stimulates platelet-derived growth factor expression in endothelial cells. Am J Physiol 1993; 265:H3-8.

62. Malek A et al. Fluid shear stress differentially modulates expression of genes encoding basic fibroblast growth factor and platelet-derived growth factor B chain in vascular endothelium. J Clin Invest 1993; 92:2013-21.

63. Resnick N et al. Platelet-derived growth factor B chain promoter contains a cis-acting fluid shear-stress-responsive element. Proc Natl Acad Sci USA 1993; 90:4591-5.

64. Uematsu M et al. Mechanisms of endothelial cell NO synthase induction by shear stress (abstract). Circulation 1993; 88:I:184.

65. Honda H et al. Disturbed flow induces heat shock protein-70 mRNA in bovine and human aortic endothelial cells (abstract). Circulation (Suppl 1) 1992; 86:I:224.

66. Malek A et al. Endothelial expression of thrombomodulin is reversibly regulated by fluid shear stress. Circ Res 1994; 74:852-60.

67. Bhagyalakshmi A et al. Fluid shear stress stimulates membrane phospholipid metabolism in cultured human endothelial cells. J Vasc Res 1992; 29:443-9.

68. Nollert M, Eskin S, McIntire L. Shear stress increases inositol triphosphate levels in human endothelial cells. Biochem Biophys Res Commun 1990; 170 281.

69. Ohno M et al. Shear stress elevates endothelial cGMP. Role of a potassium channel and G-protein coupling. Circulation 1993; 88:193-7.
70. Kuchan MJ, Frangos JA. Role of calcium and calmodulin in flow-induced nitric oxide production in endothelial cells. Am J Physiol 1994; 266:C628-36.
71. Gooch KJ, Frangos JA. Flow- and bradykinin-induced nitric oxide production by endothelial cells is independent of membrane potential. Am J Physiol 1996; 270:C546-51.
72. Prasad A et al. Flow-related responses of intracellular inositol phosphate levels in cultured aortic endothelial cells. Circ Res 1993; 72:827-36.
73. Rosales O, Sumpio B. Changes in cyclic strain increase inositol triphosphate and diacylglycerol in endothelial cells. Am J Physiol 1992; 262(31):C956-62.
74. Letsou G et al. Stimulation of adenylate cyclase activity in cultured endothelial cells subjected to cyclic stretch. J Cardiovasc Surg 1990; 31:634-9.
75. Dewey CF et al. The dynamic response of vascular endothelial cells to fluid shear stress. J Biomech Eng 1981; 103:177-88.
76. Eskin S et al. Response of cultured endothelial cells to steady flow. Microvasc Res 1984; 28:87-93.
77. Levesque M et al. Correlation of endothelial cell shape and wall shear stress in a stenosed dog aorta. Arteriosclerosis 1986; 6:220-9.
78. Levesque M, Nerem R. The study of rheological effects on vascular endothelial cells in culture. Biorheology 1989; 26:345-57.
79. Langille B, O'Donnel F. Relationship between blood flow direction and endothelial cell orientation at arterial branch sites in rabbits and mice. Circ Res 1986; 48:481-8.
80. Franke R et al. Induction of human vascular endothelial stress fibers by fluid shear stress. Nature 1984; 307:648-50.
81. Ookawa K, Sato M, Ohshima N. Changes in the microstructure of cultured porcine aortic endothelial cells in the early stage after applying a fluid-imposed shear stress. J Biomech 1992; 25:1321-8.
82. Wechezak A, Viggers R, Sauvage L. Fibronectin and F-actin redistribution in cultured endothelial cells exposed to shear stress. Lab Invest 1985; 53:639-47.
83. Wechezak A et al. Endothelial adherence under shear stress is dependent upon microfilament reorganization. J Cell Physiol 1989; 139:136-46.
84. Kim D, Gotlieb A, Langille B. In vivo modulation of endothelial F-actin microfilaments by experimental alterations in shear stress. Arteriosclerosis 1989; 9:439-45.
85. Langille B et al. Dynamics of shear-induced redistribution of F-actin in endothelial cells in vivo. Arterioscler Thromb 1991; 11:1814-20.
86. Nerem R. Vascular fluid mechanics the arterial wall and atherosclerosis. J Biomech Eng 1992; 114:274-82.

87. Nerem R et al. Hemodynamics and vascular endothelial biology. J Cardiovasc Pharmacol 1993; 21:S6-10.
88. Helmlinger G et al. Effects of pulsatile flow on cultured vascular endothelial cell morphology. J Biomech Eng 1991; 113:123.
89. Sato M, Levesque M, Nerem R. Micropipette aspiration of cultured bovine aortic endothelial cells exposed to shear stress. Arteriosclerosis 1987; 7:276-86.
90. Davies PF et al. Turbulent fluid shear stress induces vascular endothelial cell turnover in vitro. Proc Natl Acad Sci USA 1986; 83:2114-7.
91. Levesque M, Jan RM et al. Vascular endothelial cell proliferation in culture and the influence of flow. Biomaterials; 1990 11:702-7.
92. DePaola N et al. Vascular endothelium responds to fluid shear stress gradients. Arterioscler Thromb 1992; 107:1254-7.
93. Kamiya A, Togawaand T. Adapted regulation of wall shear stress to flow changes in the canine carotid artery. Am J Physiol 1980; 239:H14-21.
94. Zarins CK et al. Shear stress regulation of artery lumen diameter in experimental atherogenesis. J Vasc Surg 1987; 5:413-20.
95. Melkumyants AM et al. Continuous control of the lumen of feline conduit arteries by blood flow rate. Cardiovasc Res 1987; 21:863-70.
96. Kuchan MJ, Frangos JA. Shear stress regulates endothelin-1 release via protein kinase C and cGMP in cultured endothelial cells. Am J Physiol 1993; 264:H150-6.
97. Davidge ST et al. Nitric oxide produced by endothelial cells increases production of eicosanoids through activation of prostaglandin H synthase. Circ Res 1995; 77:274-83.
98. Morigi M et al. Supernatant of endothelial cells exposed to laminar flow inhibits mesangial cell proliferation. Am J Physiol 1993; 264(33):C1080-3.
99. Sterpetti AV et al. Shear stress modulates the proliferation rate, protein synthesis, and mitogenic activity of arterial smooth muscle cells. Surgery 1993; 113(6):691-9.
100. Ziegler T, W.E Alexander, Nerem RM. Effect of flow in the morphology and growth of endothelial cells co-cultured with smooth muscle cells. Ann Biomed Eng 1995; 23(3):216-225.
101. Frangos JA, McIntire LV, Eskin SG. Shear stress induced stimulation of mammalian cell metabolism. Biotechnol Bioeng 1988; 32:1053-60.
102. Noris M et al. Nitric oxide synthesis by cultured endothelial cells is modulated by flow conditions. Circ Res 1995; 76(4):536-43.
103. Yoshizumi M et al. Hemodynamic shear stress stimulates endothelin production by cultured endothelial cells. Biochem Biophys Res Commun 1989; 161:859-64.
104. Milner P et al. Rapid release of endothelin from isolated aortic endothelial cells exposed to increased flow. Biochem Biophys Res Commun 1990; 170:649-56.

105. Sharefkin JB et al. Fluid flow decreases preproendothelin mRNA levels and suppresses endothelin-1 release in cultured human endothelial cells. J Vasc Surg 1991; 14:1-9.

106. Kuchan MJ, Hanjoong J, Frangos J. Role of G proteins in shear stress-mediated nitric oxide production. Am J Physiol 1994; 267:C753-8.

107. Hutcheson IR, Griffith TM. Release of endothelium-derived relaxing factor is modulated both by frequency and amplitude of pulsatile flow. Am J Physiol 1991; 261:H257-62.

108. Acevedo AD et al. Morphological and proliferative responses of endothelial cells to hydrostatic pressure: role of fibroblast growth factor. J Cell Physiol 1993; 157:603-614.

109. Kato S et al. Ambient pressure stimulates immortalized human aortic endothelial cells to increase DNA synthesis and matrix metalloproteinase 1 (tissue collagenase) production. Virchows Archiv 1994; 425:385-90.

110. Hishikawa K et al. Pressure enhances endothelin-1 release from cultured human endothelial cells. Hypertension 1995; 25:449-52.

111. Mills I, Cohen CR, Sumpio BE. Cyclic strain and vascular cell biology. In: Sumpio BE, Ed. Hemodynamic Forces and Vascular Cell Biology. Austin: RG Landes, 1993:66-89.

112. Isales C, O Rosales, Sumpio BE. Mediators and mechanism of cyclic strain and shear stress-induced vascular responses. In: Sumpio BE, Ed. Hemodynamic forces and vascular cell biology. Austin: RGLandes, 1993.

113. Brophy C et al. Phospholipase C: a putative mechanotransducer for endothelial cell response to acute hemodynamic changes. Biochem Biophys Res Commun 1993; 190(2):576-81.

114. Iba T et al. Stimulation of endothelial secretion of tissue-type plasminogen activator by repetitive stretch. J Surg Res 1991; 50:457-60.

115. Iba T, Mills I, Sumpio B. Intracellular cyclic AMP levels in endothelial cells subjected to cyclic strain in vitro. J Surg Res 1992; 52:625-30.

116. Iba T, Sumpio B. Morphological response of human endothelial cells subjected to cyclic strain in vitro. Microvascular Res 1991; 42:245-54.

117. Sumpio BE, Banes AJ, Prostacyclin synthtic acticity in cultures aortic endothelial cells undergoing cyclic deformation. Surgery 1988; 104:383.

118. Wang D et al. Mechanical strain induces monocyte chemotactic protein-1 gene expression in endothelial cells. Circ Res 1995; 77:294-302.

119. Sigurdson WJ, Sachs F, Diamond SL. Mechanical perturbation of cultured human endothelial cells causes rapid increases of intracellular calcium. Am J Physiol 1993; 264:H1745-52.

CHAPTER 3

Bone Cells

Bone Structure and Function

The hardness and resiliency of bone comes from its mineralized organic matrix. This matrix is composed primarily of collagen fibers,[1] which gives bone its tensional strength. Among the fibers are proteoglycans that are believed to control the deposition of the calcium salts that hold the fibers tightly in place and provide compact bone with its compressional strength. The presence of these salts is one of the primary differences between bone and cartilage matrix, which otherwise have similar structure.

Bone matrix is continually formed by osteoblasts, which reside on the surface of bone and in cavities running through bone. Osteoblasts are formed by osteoprogenitor (stem) cells. Osteoclasts are phagocytic multinucleated cells formed in bone marrow that ordinarily absorb bone at a rate approximating that of bone deposition. This ongoing renovation of bone ensures that its strength remains constant. Osteocytes, a third type of bone cells, are differentiated osteoblasts that are trapped in the rigid bone matrix. These cells communicate with one another as well as with osteoblasts and osteoclasts through the canaliculae that run through bone matrix. Fluid forced through these canaliculae applies shear stress and generates streaming potentials adjacent to these cells, prompting speculation that they may be the primary sensors of physical stimuli in bone.

The Effects of Mechanical Forces on Bone In Vivo

The primary functions of bone are to provide a framework of support for the body and to aid in movement. Galileo observed that bone anatomy and function were related, in that body weight and physical strain placed on bones seemed to influence their size.[2] In 1867, using a curved wooden beam as an engineering model, Meyer theorized that bone density patterns were governed by the distribution of stress.[3] Julius Wolff was the first modern investigator to draw a direct correlation between mechanical stress and bone remodeling.[4] In 1892 Wolff hypothesized that bone will increase its density

Mechanical Forces: Their Effects on Cells and Tissues, by Keith J. Gooch and Christopher J. Tennant. © 1997 Landes Bioscience.

and strength in areas that are exposed to stress, while areas that are not physically stimulated will weaken and lose density. Wolff's theory that bone remodels according to physical forces is now accepted as the basis of the relationship between bone structure and function, and experiments support the contention that disuse results in bone atrophy[5-9] while physical loading increases bone mass and density.[10-14] The effects of mechanical loading are both site-specific and persistent as demonstrated by a study by Karlsson et al.[15] Individuals who underwent an extensive weightlifting regimen had greater bone mass (on the bones that were exercised) than a control group. The Karlsson experiment indicated that this bone mass discrepancy persisted for decades in some subjects. Specifically, those aged 50-64 retained greater bone mass than controls, although the weightlifting regimen of those subjects had ceased an average of 30 years earlier. However, subjects aged 65+ years exhibited bone mass identical to controls, suggesting that advanced aging counteracted the long-lasting effects of heavy exercise early in life. It is of some relevance to note that individuals of this age are most likely to suffer fragility fractures.

Interestingly, bone remodeling also can be stimulated by inducing a fluid shift within the body. In the absence of applied mechanical forces, humans subjected to head-down tilt bed rest experience a decrease in bone mass in lower extremities but an increase in bone mass in upper extremities.[16] Similar results are observed in rat tail-suspension models.[17] The mechanisms by which bone mass is affected by mechanical loading and fluid shifts within the body have never been well understood and still are the subject of investigation and speculation.

The effects of mechanical loading on bone have been correlated with the deformation (strain) the forces produce. A convenient unit of deformation is the microstrain (μstrain), with 1 μstrain equaling a 0.0001% change in length. It is believed that bone in vivo can absorb forces generating 1,500-2,500 μstrain without damage.[18] Slightly larger forces will cause microcracking, while forces generating more than 25,000 μstrain will fracture bone. Microcracking of human bones accumulates slowly under physiological conditions, but at compressive forces > 3,000 μstrain microdamage quickly accumulates,[19] causing bone resorption.[20,21] In extreme cases microcracking leads to fatigue fracture, which has been documented in many active groups, most often military recruits.[22-25] In vivo experiments utilizing loading of as little as 1,400 μstrain have been shown to engender widespread microcracking in the bones of dogs.[20] It is unclear, however, to what extent microcracking stimulates bone formation; apparently bone undergoes microcracking and remodeling continuously, and only in extreme cases will the amount of microcracking supersede the ability of the bone to repair these cracks, resulting in fracture.[26]

A hypothesis that attempts to describe the effect of small strains is the mechanostat model of bone adaptation advanced by Frost.[27] According to this model, the body attempts to maintain an optimum physiological level of bone strain, and all modeling or remodeling efforts are aimed at preserving this level.[27] For instance, 50-200 μstrain causes bone resorption rate to exceed the rate at which bone tissue is formed. This net loss of bone tissue results in an overall increase in the local strain experienced by the bone, prompting bone formation. Mechanical strains ranging from 50-2,500 μstrain are considered physiological levels in the Frost model;[28] strains of this magnitude are believed to trigger bone formation that equals bone loss, establishing an equilibrium resulting in a fixed amount of bone tissue. Strains exceeding this amount up to 4,000-5,000 μstrain are considered too great; bone formation exceeds absorption, until the mass of tissue is increased such that local strains are brought down to desirable (physiological) levels. Strains greater than 5,000 μstrain prompt excessive bone remodeling, including the formation of woven bone similar to that seen in fracture repairs. Figure 3.1 from Duncan and Turner[29] summarizes the effects of different levels of mechanical loading on bone homeostasis.

One component of Frost's model is the minimum effective strain (MES) required to elicit the biological response characterized by each of the four usage windows defined by the mechanostat model. An implication of the fairly narrow ranges of MES that separate each window is that relatively small changes in mechanical loading could be expected to have dramatic effects. A report of the bone formation rate on the endocortical surface of tibia bones subjected to four-point bending at a frequency of 2 Hz clearly illustrates different responses to small changes in loading. Applied bending loads less than 40 N did not modify bone formations, while applied loads of 52 N increased bone formation rates several-fold (Fig. 3.2).[30] Values for MES, however, tend to be variable and site-specific, suggesting that strain magnitude alone does not predict bone adaptation.

It has been observed that bone formation in vivo is not promoted by static loading[31,32] but is stimulated significantly by loading at frequencies approximating physiological levels, suggesting that the rate of strain may be an important determinant of bone remodeling.[32] Turner et al found that cyclic loading at frequencies > 0.5 Hz stimulated bone development in rats while low-frequency loading has been shown to inhibit bone formation, perhaps because the flow of blood or extracellular fluid was reduced.[32] Both the magnitude of the strain and the time required for the displacement to occur affect the strain rate. The strain rate, in turn, has been correlated with the velocity of interstitial fluid flow.[33-36] Frangos and coworkers have proposed that cellular

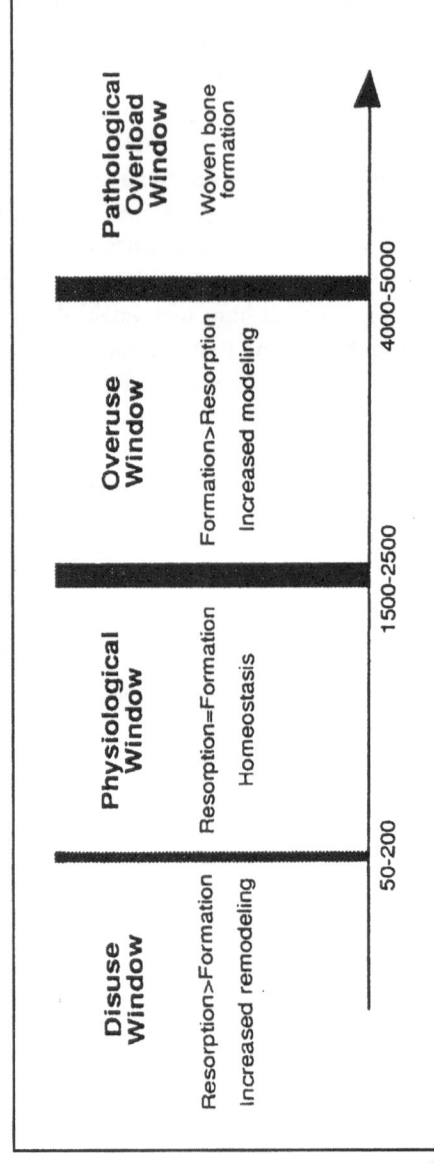

Fig 3.1. The effects of strain corresponding to the four mechanical usage windows defined by the mechanostat theory on the rate of bone resorption and deposition. Each window is separated by a minimum effective strain for which approximate values are given in microstrain. These values tend to be variable and site-specific, suggesting that strain magnitude alone does not predict bone adaptation. Duncan RL, Turner CH. Mechanotransduction and the functional response of bone to mechanical strain. With permission from Duncan RL, Turner CH. Calcif Tissue Int 1995; 57:344-58.

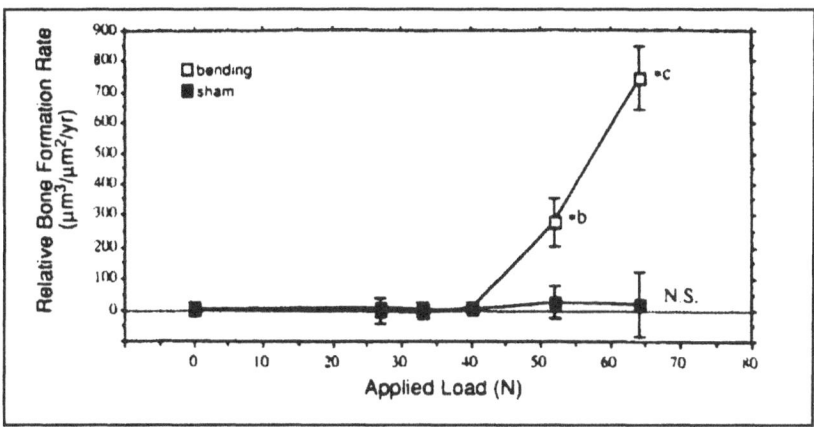

Fig. 3.2. Cyclic four-point bending at a frequency of 2 Hz was applied to the tibia of rats in vivo. Relative bone formation rate (rBFR) is the bone formation rate of the tibia subjected to bending minus that of the control. At an applied bending load of 64 N, the bone formation in the loaded tibia was six-fold greater than that of the contralateral control. With permission from Turner CH, Forwood MR, Rho J et al. Mechanical loading thresholds for lamellar and woven bone formation. J Bone Miner Res 1994; 9:87-97.

responses to shear stresses generated by mechanically induced interstitial fluid flow mediates bone remodeling,[72] the subject of a recent review.[37]

In vivo studies have been helpful in understanding the role of interstitial fluid flow in bone modeling. Montgomery et al conducted experiments in which the flow of interstitial fluid containing ferritin was measured in adult dogs.[38] They found that the movement of fluid is greatly impeded by the fibrous periosteum of bone. It is known that the periosteal layer thickens with age. It also is known that bone resorption begins to outpace bone formation as organisms age. The thickening of the periosteal layer and the resultant decrease in fluid flow suggest that flow plays a significant role in stimulating formation of new bone.[37] Interstitial fluid flow may also explain observations of increased bone mass in the skull and mandible in humans exposed to 6° head down tilt[39] or rats subjected to tail suspension.[17]

Some in vivo studies suggest that high-intensity physical stimulation stimulates the release of growth hormones,[40] elevating the amount of insulin-like growth factor-I (IGF-I) in the bloodstream, which then may act as a regulator of cellular metabolism and action.[41] Another interesting in vivo study of mechanical stimulation[42] indicated that increased prostaglandin (PG) synthesis led to an increase in glucose-6-phosphate dehydrogenase (G6PD), which is necessary to stimulate production of the ribose needed to form RNA. This is one of the few in vivo studies to trace a pathway from physical stimulation to cellular response.

The in vivo studies described above clearly demonstrate the effects of mechanical loading on bone remodeling. In vivo study has been useful in determining many short- and long-term effects of cyclic loading on bone tissue. However, the use of cell culture allows researchers to control and isolate components of force in ways that are impossible with in vivo studies. Investigators use animal cell culture as a tool to further understand which aspects of physical force account for remodeling and which biological processes are involved.

The Effects of Mechanical Forces On Bone In Vitro

Researchers are now seeking to understand the translation of physical forces to the chemical signals that result in bone remodeling. One of the advantages of in vitro studies is the level of control over variables. This control allows researchers to investigate the mechanisms by which cells convert physical stimuli into chemical messages and, ultimately, metabolic responses. A four-step system can be used to describe this conversion of physical stimuli to biological responses.[29] The first part of the process, mechanocoupling, is the conversion of a physical force (e.g., compression of a bone) to a secondary force that directly impacts the cell (e.g., deformation of cell membrane). In other words, one mechanical force is converted into another. The most common results of mechanocoupling are cell deformation and increased flow of interstitial fluid in the bone matrix, which then creates a streaming potential as well as shear stress on cells exposed to the flow. The second step, mechanotransduction, occurs when the physical force directly impacting the cell is converted into a chemical or electrical response within the cell. Researchers are investigating the role of ion channels, G proteins, and force transduction through the cytoskeleton as possible mechanisms involved in this step. Signal transduction, the next step, involves the conversion of the primary message into a secondary (and tertiary and quaternary, etc.) message. Certain second messengers, such as cyclic adenosine monophosphate (cAMP), growth factors, PG, Ca^{2+} and nitric oxide (NO) are believed to play key roles in the transduction of physical forces to chemical signals, although at present the mechanisms by which they function are poorly understood. The final step in the process, the net result of the physical stimuli, is the physiological response or functional change exhibited by the cell. For a comprehensive review of the mechanotransduction mechanisms of bone, see a review by Duncan and Turner.[29]

Cell culture studies of the effects of physical forces on bone cells generally utilize either static loading or cyclic loading, the latter of which more accurately represents the strain most bones experience in vivo. These applied strains result in two mechanical responses: cell deformation and fluid flow. Researchers have conducted experiments investigating each response over a wide range of durations, magnitudes of strain and, in cyclic strain trials, frequencies.

Static Loading

Static loading experiments are less common than cyclic loading experiments, possibly because they do not accurately recreate in vivo conditions, making their results more difficult to interpret. However, by allowing experimenters to control virtually every variable, they do illustrate certain metabolic responses. Static loading experiments in vivo have not been linked to increased bone formation,[31] but similar trials in vitro have resulted in increases in bone cell proliferation and second messenger production.[43,44] Static loading in vitro can be achieved in several ways. The most common experimental methods are increasing hydrostatic pressure (static compression) or deforming the plate on which the cells are growing (static deformation).

Static compression

In vitro studies of static compression applied to cells have indicated that supraphysiological pressure levels inhibit osteoblast differentiation,[45] destabilize the cytoskeleton,[46,47] and decrease RNA and protein synthesis.[48,49] Mononuclear mouse bone marrow cells subjected to hydrostatic pressure of 0.5-2 atm above normal atmospheric pressure exhibited greater bone resorption than controls. Researchers determined that the additional pressure inhibited differentiation of osteoblasts and stimulated generation of new osteoclasts.[45] The effect was greatest at 1 atm above normal pressure. During this time the cells showed increased production of prostaglandin E_2 (PGE_2) as well as tartrate-resistant acid phosphatase-positive mononuclear cells, possible osteoclast precursors. The addition of indomethacin negated the effects of the additional pressure on PGE_2 synthesis and mononuclear cell production,[45] a finding which was corroborated by others.[43] This suggests that PGE_2 is a primary second messenger regulating osteoclast generation.[45]

In a similar experiment, Ozawa et al[50] subjected mouse osteoblast-like cells in culture to pressures 0.5-2 atm above atmospheric pressure. The cells demonstrated an inhibited ability to synthesize collagen, and calcification and alkaline phosphatase activity were suppressed.[50] The researchers surmised that the constant pressure increased PGE_2 synthesis, which then inhibited the differentiation of osteogenic cells.

In addition to raising PGE_2 synthesis and inhibiting collagen synthesis, static pressure apparently destabilizes the cytoskeleton. Following application of a 20-min burst of hydrostatic pressure (4 MPa) to human osteosarcoma cells, Haskin et al[46] observed what seemed to be the beginning of cytoskeletal disintegration. Specifically, α-tubulin, actin and vimentin underwent significant depolymerization. The α-tubulin and vimentin chains shortened so much that no link between the nucleus and the cell membrane could be seen, in contrast to control cells. In addition, the α-tubulin and vimentin polymers no longer reached cell-cell interfaces. Actin polymers withdrew to sites within the cytoplasm and did not show significant reorganization within 2 hours following the pressure pulse.

The reduced viability of cytoskeletal structure in the previous example suggests that lines of communication between the cell periphery and the nucleus had been compromised. But the cells also exhibited increased levels of heterotrophic adhesion receptor, which functions as a cellular adhesive, at desmosomal junctions.[51,52] These researchers concluded that the cells were responding to cytoskeletal disruption by increasing adhesion to one another and to the extracellular matrix. It should be noted, however, that these pressures are likely much larger than those experienced in vivo.

Static deformation

A common method of applying static deformation to bone cells involves plating them in a monolayer on flexible plastic membranes that can be bent and held in that position. Several studies have been done in which murine or rat calvarial cells were plated on a flexible membrane and stretched for a fixed period of time.[43,53-56] Investigators have found that deformation of these cells generally results in increased levels of PG,[43,53,54] which is believed to be an important regulator of bone remodeling and growth.[43,53,54] Researchers also have noted increases in the levels of other intracellular second messengers that seem to correspond to these rises in PG level. Applying static deformation to rat embryonal periosteum and calvarial cells, researchers reported that an immediate increase in PGE_2 was followed by temporary increases in intracellular cAMP and Ca^{2+} levels (Fig. 3.3).[43] Several relationships between these levels have been forwarded; as they are somewhat contradictory, a thorough reading of the pertinent literature is recommended. Again it should be noted that the extent of deformation these cells are subjected to in vitro (~10%) is much larger than they would experience in vivo (~0.1%).

Some researchers have obtained results that differ from those discussed above. For instance, Sandy et al[55] reported that static deformation of murine calvarial osteoblasts in monolayer culture did not affect PG levels in 11 of 12 trials. The same group reported a major rise in the levels of both low- and high-molecular-weight factors that stimulate bone resorption.

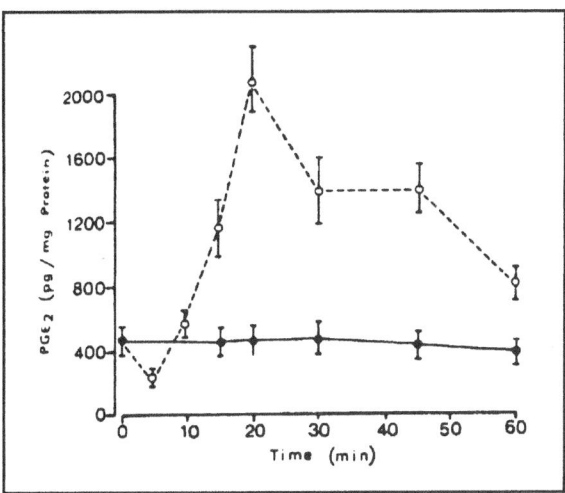

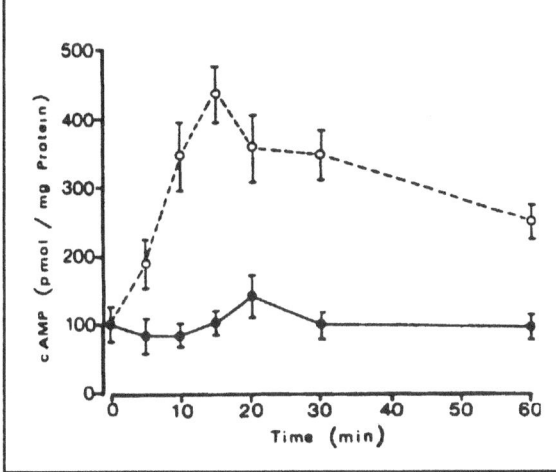

Fig. 3.3. Static stretch of bone cells results in increases in extracellular PGE$_2$ (Panel A, top) and cAMP (Panel B, bottom). The results are mean ± standard error of the mean from 3 to 6 experiments. Dotted lines correspond to strained cells and solid line corresponds to stationary controls. Somjen D, Binderman I, Berger E et al. Bone remodelling induced by physical stress is prostaglandin E2 mediated. With permission from Somjen D et al. Biochem Biophys Acta 627:91-100;1980.

Cyclic Loading

The majority of loading experienced by bone in vivo is cyclic, and as a result in vitro cyclic compression studies are more common than static loading experiments. Researchers primarily utilize two methods: cyclic compression/decompression of a chamber, and cyclic deformation of a flexible plate on which cells are growing. An important side effect of cyclic loading is the increase in fluid flow through the bone matrix and the subsequent generation of streaming potentials and shear stresses. Experiments utilizing flow chambers to investigate these phenomena will be addressed following descriptions of cyclic compression and deformation studies.

Cyclic compression

One method of exposing bone cells to cyclic compression is intermittent compression of the gas phase inside a culture chamber.[57] A series of experiments of this type conducted by Klein-Nulend et al on mouse calvarial cells in organ culture have been shown to increase tissue formation[57,58] and reduce bone resorption.[57-59] Applying cyclic compression to murine calvarial cells apparently stimulates bone formation and inhibits bone absorption by affecting the production of an autocrine growth factor and a paracrine factor that inhibits growth or differentiation of osteoblast precursor-like cells. Treatment of cells with conditioned medium from cultures subjected to cyclic compression showed a more pronounced effect than did cultures subjected directly to the compression, indicating that enhanced production of some factor was more likely responsible than the force itself.

Cyclic deformation

The cyclic deformation of cell culture is ordinarily achieved by plating cells on a flexible dish, then mechanically deforming the plate by application of force or suction, resulting in supraphysiological deformations of ~1-10%. One of the most common results of exposing bone cells to cyclic deformation is an increase in PG synthesis.[18,55,60,61] For instance, cells exposed to several 15 min periods of stretching exhibited a rate of PG synthesis more than three times that of a control group.[60] Osteoblast-like cells subjected to cyclic stretching generally have been shown to increase synthesis of both collagen and noncollagen proteins.[18,61] Trials using chick calvarial cells showed increases in both types of proteins,[18] while human osteosarcoma cells exposed to cyclic deformation (0.05 Hz) showed increased levels of type I collagen as well as osteocalcin and RNA for osteopontin, two noncollagenous bone matrix proteins associated with bone mineralization.[62-64] These noncollagenous proteins appeared despite the absence of 1,25-dihydroxyvitamin D, formerly believed to be necessary for their formation. The addition of vitamin D to the stretched cultures increased the levels of these proteins. This suggests that cyclic mechanical strain upregulates the formation of noncollagenous bone matrix proteins independent of vitamin D, their normal regulator.

Using two-dimensional gel electrophoresis, Buckley et al quantified protein synthesis by osteoblasts exposed to stationary or cyclic deformation (24% maximum strain, 0.05 Hz). Collagenous and noncollagenous protein production was enhanced while production of more than 100 other proteins simultaneously decreased (Table 3.1).[18] One protein whose synthesis increased which is of particular interest is alkaline phosphatase, an established phenotypic marker associated with bone formation and thought to play a role in mineralization (Fig. 3.4). This result suggests that mechanical forces may promote osteoblast differentiation, thereby favoring bone formation. These researchers theorized that the net result of cyclic deformation is a decrease in the production of luxury proteins with an increase in cytoskeletal and focal adhesion proteins, including vimentin, actin and tubulin.

Cyclic stretching also has been reported by some groups to stimulate proliferation[65] and DNA synthesis in bone cells. Hasegawa et al[43,56] reported a 64% increase in the number of cells synthesizing DNA after a 2-h period of deformation.[56] The experimental cells also had a significantly higher rate of protein synthesis as measured by incorporation of ^{3}H-proline and -leucine, although collagen synthesis rates were the same for experimental and control groups.

Table 3.1. Synthesis of known proteins affected by cyclic deformation.

Protein	Molecular weight (kDa)	Isoelectric point (pI)	Protein synthesis
			(fold change)
Calmodulin	17	4	2.12
α-Tubulin	57.4	5.21	8.7
Vimentin	61	5.16	22
HSP 80	80	5.19	0.77
HSP 90	90	5.19	↓
HSP 100	100	5.09	0.063
Actin	42	5.42	0.50
HSP 73	73	5.5	↑
Vinculin	130	6.07	5.6
Phosphorylase B	20	6.96	18

Protein synthesis by osteoblasts exposed to cyclic strain (24% maximum strain, 0.05 Hz) or stationary conditions were quantified using two-dimensional gel electrophoresis. ↑ indicates an increase from undetectable to detectable levels and ↓ indicates a decrease from detectable to undetectable levels. Modified from ref. 18.

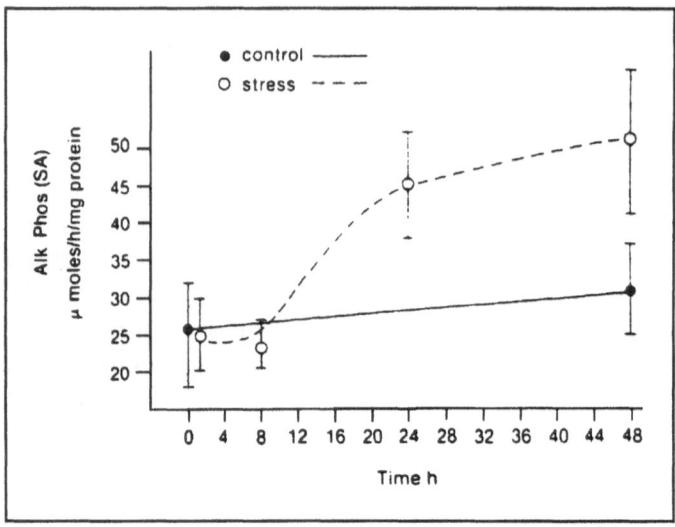

Fig 3.4. Alkaline phosphatase activitity of cultured osteoblast exposed to cyclic strain (24% maximum strain, 0.05 Hz) or stationary conditions for 24 hours. With permission from Buckley MJ, Banes AJ, Jordan RD. The effects of mechanical strain on osteoblasts in vitro. With permission from Buckley MJ et al. J Oral Maxillofacial Surg 48:276-282;1990.

Exposure of chick calvarial cells to cyclic deformation (0.05 Hz) by Buckley et al prompted a striking response. In addition to an increase in cell division, the cells were observed to align perpendicularly to the applied force (Fig. 3.5, Panels A and B).[65] Realignment began at the edges of the plates, where deformation was greatest. Cells aligned within 4 h after initiation of strain, and realignment was significant within 12 h. An increase in the amount of filamentous (F) actin was seen concurrently; the investigators hypothesized that the cells responded to the deformation by releasing their focal contacts and migrating so as to minimize strain.

Osteoblast-like osteosarcoma cells exposed to cyclic strain (0.05 Hz) also have been shown to have 3-5 times the number of open stretch-activated cation channels,[66] markedly increasing whole cell conductance. This increase was negated by the addition of Gd^{3+}, a known inhibitor of such channels. The response of these channels to cyclic stress suggests that they may be of importance in mechanotransduction.

Fluid Flow

A subject of much speculation is whether cyclic compression of cells acts through physical deformation of the cells or through compression-induced fluid flow. The mineralized matrix of bone lends the structure a rigidity that generally prevents compression.[67] Under physiological conditions, bone cells experience unidirectional strain that does not exceed 0.4%, whereas muscle cells in the body can be subjected to stretching in excess of 50%.[68] This suggests that, at least for bone cells, a mechanism other than cell deformation may exist that allows tissue to sense and respond to compression.

It has been shown that compression of bone tissue forces fluid through the trabecular network within the central cavity of bones[69] and through the canaliculae of bone.[38,69,70] These canaliculae also house the minute cell processes linking the osteocytes that are bound inside the matrix. Several researchers have presented data indicating that fluid flow, rather than cell deformation, is the factor prompting the cellular responses outlined above. Fluid has been shown to increase synthesis of second messengers such as PGE_2,[69,71] intracellular $cAMP$[72] and inositol triphosphate (IP_3).[69] Fluid flow affects cells in two ways. The tangential force of fluid flowing over the cell is referred to as shear stress; the electrical potential generated by this ion-containing fluid is known as the streaming potential.

Shear stress

Researchers have obtained evidence that shear stress is responsible for at least some, and perhaps many, flow-dependent metabolic changes in bone cells.[69,72] Altering the viscosity of the perfusion medium allows researchers to differentiate effects caused by streaming potential from those caused by shear stress. Using this method, investigators have verified that the responses studied in bone cells were stimulated by shear stress rather than streaming potentials.[72,73] A series of experiments by Reich et al indicate that production of the second messengers $cAMP$ and PGE_2 and IP_3[69] are increased in osteoblasts subjected to shear stress (Fig. 3.6). Rat calvarial osteoblasts exhibited a nine-fold increase in PGE_2 production when exposed to shear of 6 dyn/cm^2. Exposure of the cells to 24 dyn/cm^2 of shear resulted in a 20-fold increase in PGE_2 as well as a significant increase in the level of IP_3,[69] which was not seen in the group exposed to lower shear levels. In similar flow experiments, $cAMP$ concentration was 12 times greater than stationary controls.

Similar trials conducted with chick calvarial cells showed elevated PG production in osteocytes (after 6 and 24 h) and in osteoblasts (after 6 h).[16] It is believed that autocrine or paracrine factors regulating bone growth are synthesized by bone cells,[17,18] and researchers have been able to trigger their synthesis with cyclic compression. Chick calvarial osteocytes subjected to 1 hr of pulsating fluid flow showed elevated levels of PGE_2, but osteoblasts and periosteal fibroblasts subjected to the same conditions showed no change. This suggests that in bone, osteocytes may be more sensitive to fluid flow and may play an important role in converting flow to chemical signals within bone tissue.

Streaming potential

In 1957 researchers observed electrical currents caused by the compression of dry bone.[76] Further investigation[77-79] showed that this piezoelectric effect in wet bone was vastly superseded by streaming potentials, which are caused by the flow of ion-containing fluid over a charged surface.

Streaming potentials set up in bone cell culture have not stimulated bone cell proliferation or activity.[80] The major effect of streaming potentials may be their ability to draw osteoclasts and osteoblasts to different areas within bone matrix (Fig. 3.7). Bending of a bone results in a net positive charge on the convex surface. It has been shown experimentally that osteoclasts migrate to a positive electrode[81] while osteoblasts move toward a negative electrode.[81] This suggests that osteoclasts would be drawn to the side of a bone undergoing tensile force while osteoblasts would tend to collect in areas subjected to compressional forces, a hypothesis that has been borne out in the

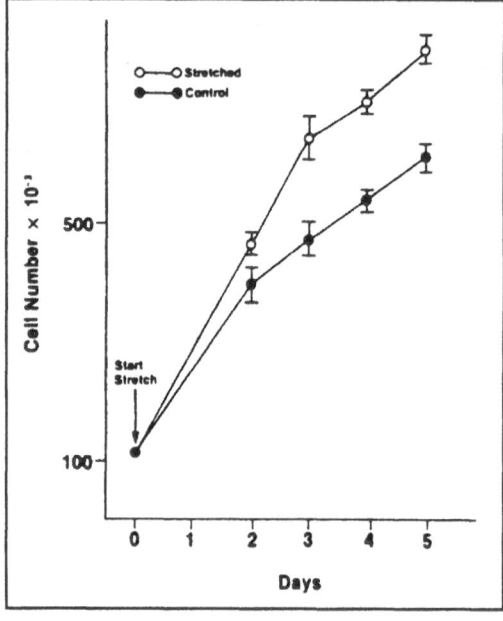

Fig. 3.5A. Results of the growth curve experiment indicated that chick calvarial osteoblasts were stimulated to divide by exposure to applied cyclic deformation to the culture plate substratum. These results represent data from one of three similar experiments with n=6/time point. Open circles represent the curve for control cells that were not stretched. Each point is the mean, ±1 standard deviation, of 6 replicate cell counts performed using a Coulter cell counter. Cell numbers were significantly greater in the stretched cells from days 2 to 5. The stimulatory event occurred during the first 72 h after initiation of cyclic deformation, yielding a 1.5-fold increase in the growth rate. After 72 h the rates were the same for each group. Reprinted by permission of the publisher from Buckley MJ et al. Osteoblasts increase their rate of division and align in response to cyclic, mechanical tension in vitro. Bone and Mineral 4(1):225-36. Copyright 1988 by Elsevier Science Inc.

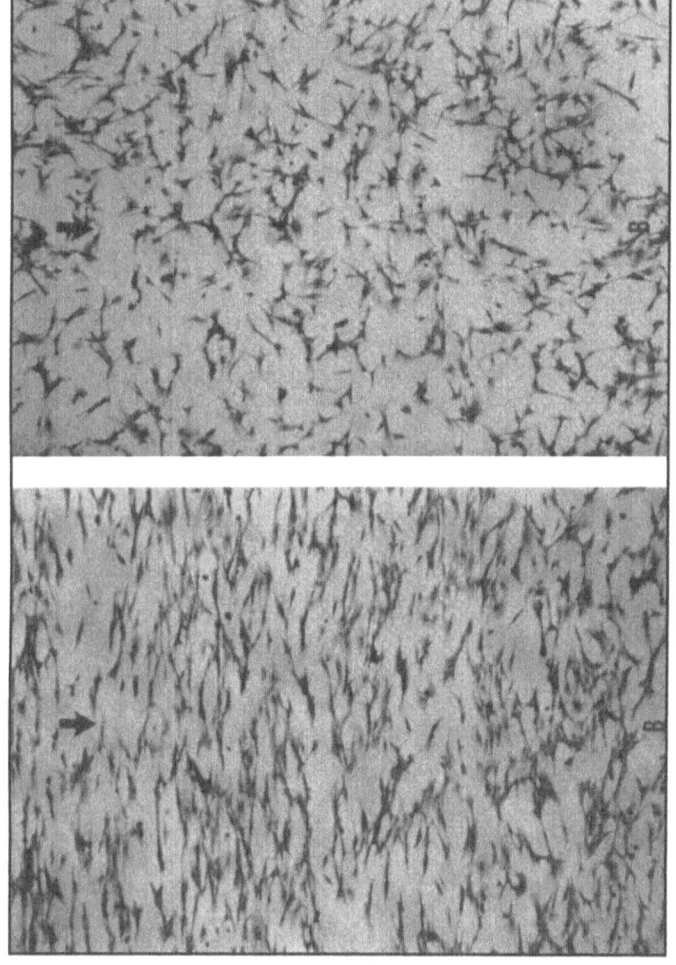

Fig. 3.5B. Photograph of chick osteoblast that were plated at 100,000 cells per well of a six well Flex I® plate and photographed 48 h later (25 h after initiation of the cyclic deformation regimen). After 24 h of cyclic deformation, osteoblasts were uniformly aligned perpendicular to the direction of the strain field, in an annular fashion, at the periphery of the culture dish. The arrow indicates the direction of the strain field and points from the perimeter toward the center of the well. Cells aligned as early as 4 h after initiation of cyclic strain. The alignment response was observed in every growth curve experiment. (Right) represents a 12.5X magnification of a photograph of control chick calvarial osteoblasts stained with crystal violet. Cells were randomly oriented in the control cultures. The arrow is located at the perimeter of the culture well and points toward the well center. Reprinted by permission of the publisher from Buckley MJ et al. Osteoblasts increase their rate of division and align in response to cyclic, mechanical tension in vitro. Bone and Mineral 4(1):225-36. Copyright 1988 by Elsevier Science Inc.

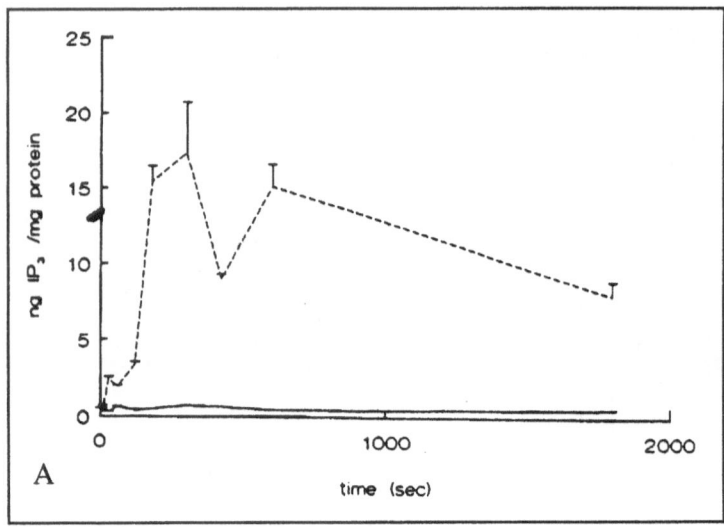

Fig 3.6. Laminar fluid flow stimulates production of IP$_3$ (Panel A, this page) and PGE$_2$ (Panel B, opposite page) by cultured osteoblasts. With permission from Reich KM, Frangos JA. Effect of flow on prostaglandin E2 and inositol triphosphate levels in osteoblasts. Am J Physiol 1991; 261:C428-32.

laboratory.[82] Another possible effect of streaming potentials on bone tissue could be their effect on voltage-gated channels,[26] but data describing this effect is scarce. Similarly, exposure of living bone to electromagnetic fields is known to stimulate bone formation,[83] but no relation between streaming potentials and bone formation has been established.

Summary

Bone provides a solid structural framework for the body and aids in movement. The hardness and resiliency of bone is derived from its mineralized organic matrix, which is composed of collagen fibers and proteoglycans. This matrix has the ability to vary its density depending on the amount of mechanical loading to which it is exposed, although the mechanism by which this is accomplished is not well understood.

One hypothesis regarding mechanically induced remodeling of bone is that the body attempts to maintain an optimum level of bone strain, and all modeling and remodeling efforts are aimed at preserving this level. Bone strains ranging from 50-2,500 µstrain (1 µstrain = 0.0001% change in bone length) are considered physiologically appropriate; at these levels of strain bone formation roughly equals resorption. Accordingly, strain below this level results in bone atrophy, while exposure to greater levels results in increased

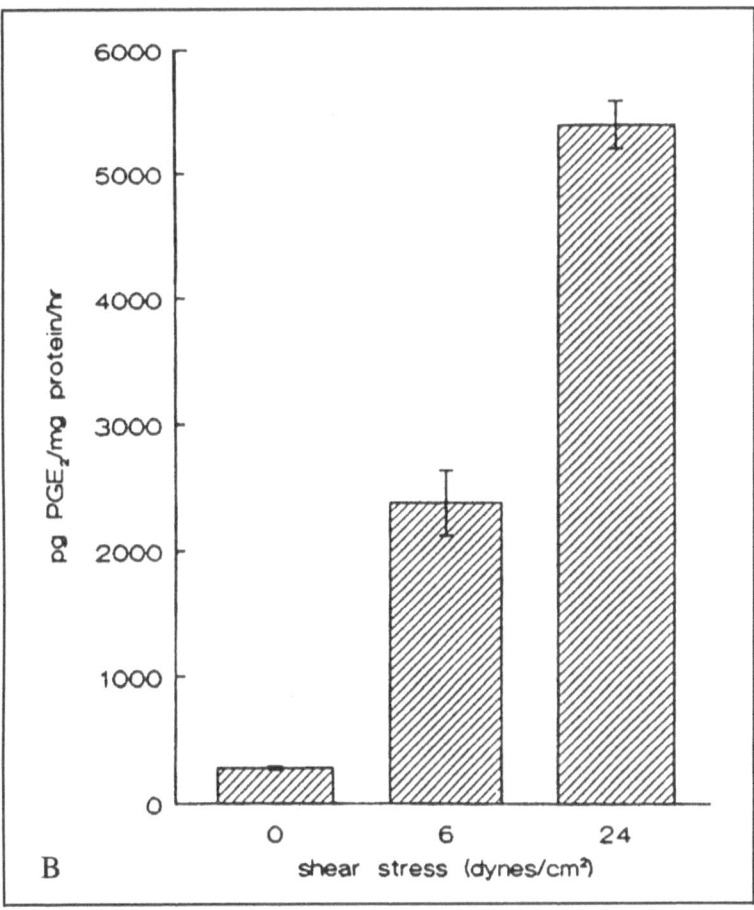

Fig. 3.6. (Panel B, see legend, opposite page).

bone deposition. Strains exceeding 5,000 µstrain can cause excessive bone deposition, including the formation of woven bone. The values of strain required for each mode of bone remodeling tend to be variable and site-specific, suggesting that strain magnitude alone does not predict bone adaptation. Growing evidence suggests that the strain rate, or even the fluid flow it generates, is the mechanical signal to which the cells respond.

In vitro studies of bone cell responses to mechanical forces generally utilize either static or cyclic strain, pressure, or fluid flow. Static pressure experiments in vitro have resulted in increased bone cell proliferation and second messenger production. Static deformation experiments, usually run on cells plated on flexible surfaces, generally have resulted in increased levels of second messengers and prostaglandins.

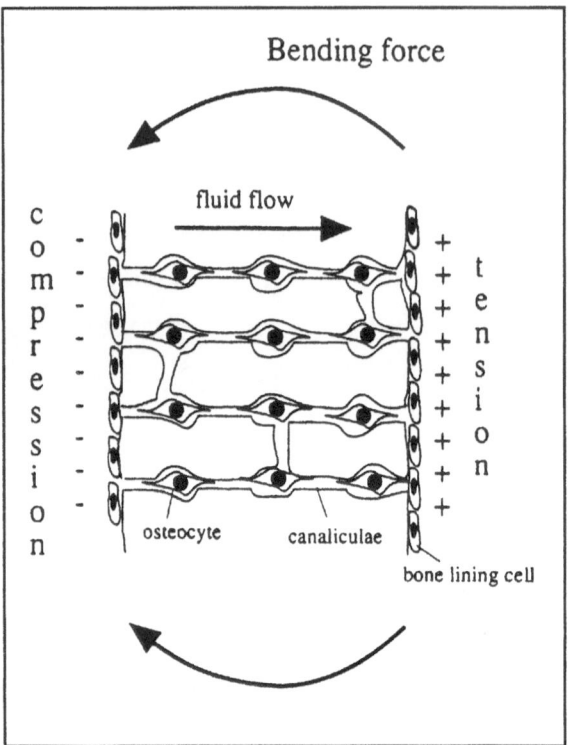

Fig 3.7 Deformation of bone induces a potential gradient which may in turn result in osteoblast and osteoclast migration. With permission Duncan RL, Turner CH. Mechanotransduction and the functional response of bone to mechanical strain. Calcif Tissue Int 1995; 57:344-58.

More commonly, researchers expose bone tissue to cyclic loading, which is more analogous in kind to physiological strain than static loading, though the magnitude of strain often employed in in vivo studies (e.g., >10%) is orders of magnitude greater than that a bone cell might experience in vivo (e.g., ~0.1%). Cyclic strain of osteoblasts has been shown to increase prostaglandin synthesis and affect protein synthesis.

Both pulsatile and steady laminar fluid flow stimulate osteoblast production of nitric oxide and prostaglandins, two important modulators of bone homeostasis. These results, and in vivo studies corroborating the role of interstitial fluid flow as a stimulant of net bone formation, suggest that applied mechanical loads may result in interstitial fluid flow, which in turn stimulates bone cells.

References

1. Guyton AC. Textbook of Medical Physiology. 8th ed. Philadelphia: WB Saunders, 1991:872-4.
2. Galilei G. Discorsi e dimonstrzioni mathematiche intorna a due nuove scieze. In: Treharne RW. Review of Wolffs Law and its proposed means of operation. Orthopaed Rev 1981; 10:35.
3. Meyer GH. Die Architektur der Spongiosa. Arch Anat Physiol Wiss Med 1867; 34:615-28.
4. Wolff J. Das Gesetz der Transformation der Knochen. Berlin: Hirschwald, 1892.
5. Jaworski ZFG, Uhthoff HK. Reversibility of nontraumatic disuse osteoporosis during its active phase. Bone 1986; 7:431-9.
6. Li XJ, Jee WSS, Chow S-Y et al. Adaptation of cancellous bone to aging and immobilization in the rat: a single photon absorptiometry and histomorphometry study. Anat Rec 1990; 227:12-24.
7. Mack PB, LaChance PA. Effects of recumbency and space flight on bone density. Am J Clin Nutr 1967; 20:1194.
8. Morey ER, Baylink DJ. Inhibition of bone formation during space flight. Science 1978; 201:1138.
9. Donaldson CL, Hulley SB, Vogel JM et al. Effect of prolonged bed rest on bone mineral. Metabolism 1970; 19:1071-84.
10. Burr DB, Schaffler MB, Yang KH et al. Skeletal change in response to altered strain environments: is woven bone a response to elevated strain? Bone 1989; 10:223-33.
11. Hert J, Liskova M, Landrgot B. Influence of the long-term continuous bending on the bone: an experimental study on the tibia of a rabbit. Folia Morphol (Prague) 1969; 27:389-99.
12. Lanyon LE, Goodship AE, Pye CJ et al. Mechanically adaptive bone re-modeling. J Biomech 1982; 15:141-54.
13. Turner CH, Forwood MR, Rho J et al. Mechanical loading thresholds for lamellar and woven bone formation. J Bone Miner Res 1994; 9:87-97.
14. Rubin CT, Lanyon LE. Regulation of bone mass by mechanical strain magnitude. Calcif Tissue Int 1985; 37:411-417.
15. Karlsson MK, Johnell O, Obrant KJ. Is bone mineral density advantage maintained long-term in previous weight lifters? Calcif Tissue Int 1995; 57:325-8.
16. Arrand S. Aviation Space Env Med 1992, 14-20.
17. Roer and Dillman J. Bone growth and calcium loss during simulated weightlessness in the rat. Appl Physiol 1990; 68:13-20.
18. Buckley MJ, Banes AJ, Jordan RD. The effects of mechanical strain on osteoblasts in vitro. J Oral Maxillofac Surg 1990; 48:276-82.
19. Carter DR, Caler WE. A cumulative damage model for bone fracture. J Orthop Res 1985; 3:84-90.
20. Burr DB, Martin RB, Schaffler MB et al. Bone remodelling in response to in vivo fatigue microdamage. J Biomech 1985; 18:189-200.

21. Mori S, Burr DB. Increased intracortical remodeling following fatigue damage. Bone 1993; 14:103-9.
22. Bargren JH, Tilson DH. Prevention of displaced fatigue fractures of the femur. J Bone Joint Surg 1971; 53A:1115, 1971.
23. Krause GR, Thompson JR Jr. March fracture: an analysis of two hundred cases. Am J Roentgenol Radium Ther Nucl Med 1944; 52:281.
24. Gilbert RS, Johnson HA. Stress fractures in military recruits: a review of twelve years experience. Milit Med 1966; 131:716.
25. Morris JM, Blickenstaff LD. Fatigue Fractures. Springfield: Charles C Thomas, 1967.
26. Carter DR, Hayes WC. Compact bone fatigue damage: a microscopic examination. Clin Orthop 1977; 127:265-74.
27. Frost HM. The mechanostat: a proposed pathogenic mechanism of osteoporoses and the bone mass effects of mechanical and nonmechanical agents. Bone Miner 1987; 2:73-85.
28. Burr DB, Martin RB. Mechanisms of bone adaptation to the mechanical environment. Triangle: Sandoz J Med Sci 1992; 31:59-76.
29. Duncan RL, Turner CH. Mechanotransduction and the functional response of bone to mechanical strain. Calcif Tissue Int 1995; 57:344-58.
30. Turner CH, Forwood MR, Rho J et al. Mechanical loading thresholds for lamellar and woven bone formation. J Bone Min Res 1994;9:87-97.
31. Lanyon LE, Rubin CT. Static versus dynamic loads as an influence on bone remodeling. J Biomech 1984; 17:897-905.
32. Turner CH, Forwood MR, Otter MW. Mechanotransduction in bone: do bone cells act as sensors of fluid flow? FASEB J 1994; 8:875-8.
33. Otter MW, Shoenung J, Williams WS. Evidence for different sources of stress-generated potentials in wet and dry bone. J Orthop Res 1985; 3:321-4.
34. Salzstein RA, Pollack SR. Electromechanical potentials in cortical bone. II. Experimental analysis. J Biomech 1987; 20:271-80.
35. Scott GC, Korostoff E. Oscillatory and step response electromechanical phenomena in human and bovine bone. J Biomech 1990; 23:127-43.
36. Otter MW, Palmieri VR, Wu DD et al. A comparative analysis of streaming potentials in vivo and in vitro. J Orthop Res 1992; 10:710-9.
37. Hilsley MV, Frangos JA. Review: Bone tissue engineering: the role of interstitial fluid flow. Biotech Bioeng 1994; 43:573-81.
38. Montgomery RJ, Sutker BD, Bronk JT et al. Interstitial fluid flow in cortical bone. Microvasc Res 1988; 35:295-307.
39. Arnaud SB, Sherrard DJ, Maloney N et al. Effects of 1-week head-down tilt bed rest on bone formation and the calcium endocrine system. Aviat Space Environ Med 1992; 63:14-20.
40. Felsing NE, Brasel J, Cooper DM. Effect of low- and high-intensity exercise on circulating growth hormone in men. J Clin Endocrinol Metab 1992; 75:157-62.
41. Maiter D, Underwood LE, Maes M et al. Different effects of intermittent and continuous growth hormone (GH) administration on serum

somatomedin-C/insulin-like growth factor I and liver GH receptors in hypophysectomised rats. Endocrinology 1988; 123:1053-9.

42. Rawlinson SCF, Mohan S, Baylink DJ et al. Exogenous prostacyclin, but not prostaglandin E2, produced similar responses in both G6PD activity and RNA production as mechanical loading, and increases IGF-II release, in adult cancellous bone in culture. Calcif Tissue Int 1993; 53:324-9.

43. Somjen D, Binderman I, Berger E et al. Bone remodelling induced by physical stress is prostaglandin E2 mediated. Biochim Biophys Acta 1980; 627:91-100.

44. Binderman I, Shimshoni Z, Somjen D. Biochemical pathways involved in the translation of physical stimulus into biological message. Calcif Tissue Int 1984; 36:S82-5.

45. Imamura K, Ozawa H, Hiraide T et al. Continuously applied compressive pressure induces bone resorption by a mechanism involving prostaglandin E2 synthesis. J Cell Physiol 1990; 144:222-8.

46. Haskin CH, Cameron I. Physiological levels of hydrostatic pressure alter morphology and organization of cytoskeletal and adhesion proteins in MG-63 osteosarcoma cells. Biochem Cell Biol 1993; 71:27-35.

47. Haskin CL, Athanasiou KA, Klebe R et al. A heat-shock-like response with cytoskeletal disruption occurs following hydrostatic pressure in MG-63 osteosarcoma cells. Biochem Cell Biol 1993; 71:361-71.

48. Bourns B, Franklin S, Cassimeris L et al. High hydrostatic pressure effects in vivo: changes in cell morphology, microtubule assembly and actin organization. Cell Motility Cytoskeleton 1988; 10:380-90.

49. Swezey RR, Somero GN. Pressure effects on actin self assembly: interspecific differences in the equilibrium and kinetics of the G to F transformation. Biochemistry 1985; 24:852-60.

50. Ozawa H, Imamura K, Abe E et al. Effect of a continuously applied compressive pressure on mouse osteoblast-like cells (MC3T3-E1) in vitro. J Cell Physiol 1990; 142:177-85.

51. Wayner EA, Carter WG. Identification of multiple cell adhesion receptors for collagen and fibronectin in human fibrosarcoma cells possessing unique α and common β subunits. J Cell Biol 1987; 105:1873-84.

52. Gallatin WM, Wayner EA, Hoffman PA et al. Structural homology between lymphocyte receptors for high endothelium and class II extracellular matrix receptor. Proc Natl Acad Sci USA 1989; 86:4654-8.

53. Harell A, Dekel S, Binderman I. Biochemical effect of mechanical stress on cultured bone cells. Calcif Tissue Res 1977; 22(S):202-7.

54. Binderman I, Zor U, Kaye AM et al. The transduction of mechanical force into biochemical events in bone cells may involve activation of phospholipase A2. Calcif Tissue Int 1988; 42:261-6.

55. Sandy JR, Meghji S, Scutt AM et al. Murine osteoblasts release bone-resorbing factors of high and low molecular weights: stimulation by mechanical deformation. Bone Miner 1989; 5:155-68.

56. Hasegawa S, Sato S, Saito S et al. Mechanical stretching increases the

number of cultured bone cells synthesizing DNA and alters their pattern of protein synthesis. Calcif Tissue Int 1985; 37:431-6.

57. Klein-Nulend JK, Semeins CM, Veldhuijzen JP et al. Effect of mechanical stimulation on the production of soluble bone factors in cultured fetal mouse calvariae. Cell Tissue Res 1993; 271:513-7.

58. Klein-Nulend J, Veldhuijzen JP, de Jong M et al. Increased bone formation and decreased bone resorption in fetal mouse calvaria as a result of intermittent compressive force in vitro. Bone Miner 1987; 2:441-8.

59. Klein-Nulend J, Veldhuijzen JP, van Strien ME et al. Inhibition of osteoclastic bone resorption by mechanical stimulation in vitro. Arthritis Rheum 1990; 33:66-72.

60. Yeh C-K, Rodan GA. Tensile forces enhance prostaglandin E synthesis in osteoblastic cells grown on collagen ribbons. Calcif Tissue Int 1984; 36:S67-71.

61. Harter LV, Hruska KA, Duncan RL. Human osteoblast-like cells respond to mechanical strain with increased bone matrix protein production independent of hormonal regulation. Endocrinology 1995; 136:528-35.

62. Ibaraki K, Whitson JD, Termine SW et al. Bone matrix mRNA expression in differentiating fetal bovine osteoblasts. J Bone Miner Res 1992; 743-54.

63. Owen TA, Aronow MS, Barone LM et al. Pleiotropic effects of vitamin D on osteoblast gene expression are related to the proliferative and differentiated state of the bone cell phenotype: dependency upon basal levels of gene expression, duration of exposure and bone matrix competency in normal rat osteoblast cultures. Endocrinology 1991; 128:1496-1504.

64. Malone JD, Teitelbaum SL, Griffin RM et al. Recruitment of osteoclast precursors by purified bone matrix constituents. J Cell Biol 1982; 92:227-30.

65. Buckley MJ, Banes AJ, Levin LG et al. Osteoblasts increase their rate of division and align in response to cyclic, mechanical tension in vitro. Bone Miner 1988; 4:225-36.

66. Duncan RL, Hruska KA. Chronic, intermittent loading alters mechanosensitive channel characteristics in osteoblast-like cells. Am J Physiol 1994; 36:F909-16.

67. Cowin SC, Moss-Salentjin L, Moss ML. Candidates for the mechanosensory system in bone. Adv Bioengin 1991; 20:313-6.

68. Rubin CT, Lanyon LE. Limb mechanics as a function of speed and gait: a study of functional strains in the radius and tibia of horse and dog. J Exp Biol 1982; 101:187-211.

69. Reich KM, Frangos JA. Effect of flow on prostaglandin E2 and inositol triphosphate levels in osteoblasts. Am J Physiol 1991; 261:C428-32.

70. Kufahl RH, Saha S. A theoretical model for stress-generated flow in the canaliculi-lacunae network in bone tissue. J Biomech 1990; 23:171-80.

71. Reich KM, Frangos JA. Protein kinase C mediates flow-induced prostaglandin E2 production in osteoblasts. Calcif Tissue Int 1993; 52:62-66.
72. Reich KM, Gay CV, Frangos JA. Fluid shear stress as a mediator of osteoblast cyclic adenosine monophosphate production. J Cell Physiol 1990; 143:100-4.
73. Klein-Nulend J, van der Plas A, Semeins CM et al. Sensitivity of osteocytes to biomechanical stress in vitro. FASEB J 1995; 9:441-5.
74. Canalis E, Centrella M. Isolation of a nontransforming bone derived growth factor from medium conditioned by fetal rat calvariae. Endocrinology 1986; 118:2002-8.
75. Mohan S, Linkhart T, Jennings J et al. Chemical and biological characterization of low molecular weight skeletal growth factor from human bones. Biochim Biophys Acta 1986; 884:234-42.
76. Fukada E, Yasuda I. J Phys Soc Jpn 1957; 12:1158-62.
77. Anderson JC, Erikson C. Electrical properties of wet collagen. Nature 1968; 218:166-8.
78. Gross D, Williams WS. Streaming potential and the electromechanical response of physiologically moist bone. J Biomech 1982; 15:277-95.
79. Pollack SR, Salzstein R, Pienkowski D. Streaming potentials in fluid-filled bone. Ferroelectrics 1984; 60:297-309.
80. Harrington DB, Meyer R. Effects of small amounts of electric current at the cellular level. Ann NY Acad Sci 1979; 283:301-6.
81. Ferrier J, Ross SM, Kanehisa J et al. Osteoclasts and osteoblasts migrate in opposite directions in response to a constant electrical field. J Cell Physiol 1986; 129:283-8.
82. Brighton CT, McCluskey WP. Cellular response and mechanisms of action of electrically induced osteogenesis. Bone Miner Res 1986; 4:213-54.
83. McLeod KJ, Rubin CT. The effect of low-frequency electrical fields on osteogenesis. J Bone Joint Surg 1992; 74A:920-9.

CHAPTER 4

Chondrocytes

Cartilage Structure and Function

Articular cartilage provides an impact-resistant, low-friction covering for the surface of joints. It is composed mainly of water, which is nearly incompressible, allowing it to support some of the heaviest loads in the body while remaining flexible. As such, cartilage must transmit and distribute forces generated by body weight and motion to the underlying bone. These forces are of considerable magnitude; the pressure exerted on articular cartilage in the hip while standing is estimated to be 0.7 MPa, or approximately 7 atm.* This rough estimate compares favorably to experimental values of approximately 1 MPa measured for the human hip.[1,2] Peak pressures during loading would be expected to be much higher. The stress measured in the hip prosthesis of an elderly woman walking, for instance, cycled from atmospheric pressure (0.1 MPa) to nearly 4 MPa at a frequency of approximately 1 Hz.[3] Stresses approaching 20 MPa were measured when the same individual stood up from a chair.

Chondrocytes are the predominant cell type found in cartilage, a material which is avascular, aneural, and alymphatic. Chondrocytes compose less than 10% of cartilage volume[4] and therefore play only a minor role in directly determining the mechanical properties of cartilage; however, they are responsible for the formation, maintenance and remodeling of the extracellular matrix, a biphasic gel-like layer that gives cartilage its strength, flexibility and resilience. Chondrocytes exist as isolated cells within this matrix. The matrix itself consists of a relatively coarse network of collagen (primarily type II) fibers interspersed with massive aggregates of proteoglycan (PG) molecules. The balance of the matrix is water, which accounts for the majority

---

This number was calculated assuming the weight of an average individual is 70 kg x 9.81 m/s² (700 N) and a rough estimate of the load-bearing area of cartilage (10 cm²). 1 atm = 0.1013 x 10⁶ N/m² (MPa) = 39.4 milliosmolar @ 37° C.

Mechanical Forces: Their Effects on Cells and Tissues, by Keith J. Gooch and Christopher J. Tennant. © 1997 Landes Bioscience.

of the mass of cartilage (65-80%). Collagen fibers, usually of diameter 30-200 nm[5], compose 15-25% of matrix mass. Models of these fibers show gaps of approximately 60-200 nm.[6] This network of fibers entraps aggregates of very large PG molecules, massive polymers on the order of 10^4 kDa that compose 3-10% of matrix mass. These molecules consist of a hyaluronate backbone and 30-100 monomers, which are composed of many (~50) glycosaminogly-cans (GAG) bound to a core protein.[7] One of the most important aspects of the GAG is that they each contain many negatively charged sulfate and car-boxyl groups. These groups give the PG molecules fixed negative charges that cannot move relative to the molecule, in contrast to ions in solution, which move with the fluid. Flow of fluid through a porous medium is described by Darcy's Law. An analog of this law that includes a term accounting for os-motic pressure describes fluid flow into and out of cartilage:**

$$J = \frac{k \, (P_{swelling} - P_{applied})}{L}$$

where J is the flux (volumetric flow rate divided by area) of fluid into or out of the cartilage; k is hydraulic conductivity; and $P_{swelling}$ characterizes a substance's tendency to swell and equal the external pressure (either hydro-static or osmotic) that must be applied to prevent tissue swelling (i.e., J = 0). $P_{swelling}$ can further be defined as the osmotic pressure of cartilage minus the resistance to swelling provided by the collagen gel. The value of L is the dis-tance over which the pressure differences act. The very small diameter of the pores in and between the proteoglycans (~2 nm) makes fluid flow very diffi-cult, corresponding to a very small value for hydraulic conductivity.[8] This makes cartilage very resistant to deformation, even in the presence of large forces.

Another factor that enables cartilage to resist deformation is its high os-motic pressure. As shown in Fig. 4.1, osmotic pressure (π) increases rapidly as the density of the fixed negative charges increases. This fixed charge den-sity stems from the negatively charged sulfate and carboxyl groups compos-ing the GAG subunits on the PG. When cartilage is compressed, its volume decreases, but the number of fixed charges remains constant, leading to an increase in fixed charge density. This, in turn, causes the osmotic pressure to rise, leading to an increase in $P_{swelling}$. As $P_{swelling}$ approaches $P_{applied}$, flux of wa-ter out of the cartilage approaches zero. That is, equilibrium is reached and cartilage volume remains constant. If the pressure applied to compressed car-tilage is reduced, this process will reverse and the cartilage will return to its original shape and volume.

** *This equation combines ideas from references 8 and 9, but is not presented in either.*

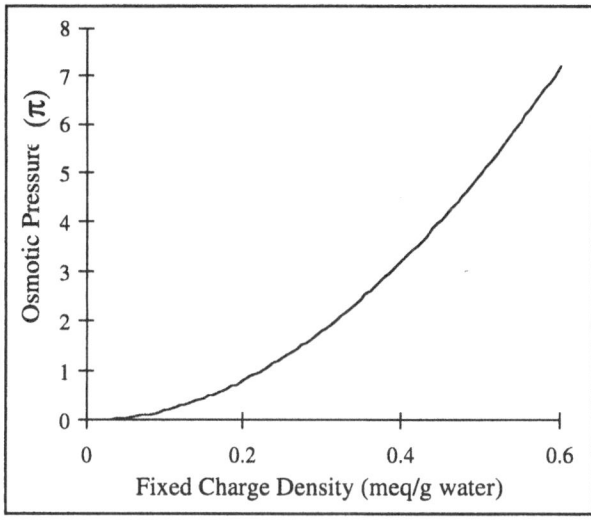

Fig. 4.1. The osmotic pressure (π) measured in atm as a function of fixed charge density for different preparations of PG with varying sizes and degree of aggregation.

PG, the primary determinant of fixed charge density, is a very large macromolecule; as such, its molar concentration is relatively low and has only minimal direct impact on osmotic pressure. However, PG and their charged groups affect osmotic pressure indirectly by altering the concentration of mobile cations and anions in the pericellular space, hence changing the osmotic pressure of this region. In order to maintain electrical neutrality, the fixed negative charges on PG attract and retain mobile cations (Na^+, K^+, Ca^{2+} and H^+) from outside the cartilage while repelling anions (Cl^-, HCO_3^-, PO_4^{2-}, SO_4^{2-} and OH^-) from within the cartilage. The presence of fixed negative charges in cartilage results in a higher concentration of mobile cations and a lower concentration of mobile anions, leading to a hyperosmotic, acidic environment in the extracellular space of cartilage (pericellular space).

This resulting redistribution of ions is described by the Gibbs-Donnan equilibrium, which states that the ratio of the concentration of an ion inside the cartilage to its concentration outside the cartilage is the same for all similarly charged ions. By coupling the Gibbs-Donnan equilibrium with the constraints of electric neutrality both inside and outside the cartilage, the concentration of cations and anions inside the cartilage is derived as a function of fixed charge density and external cation concentration. In addition,

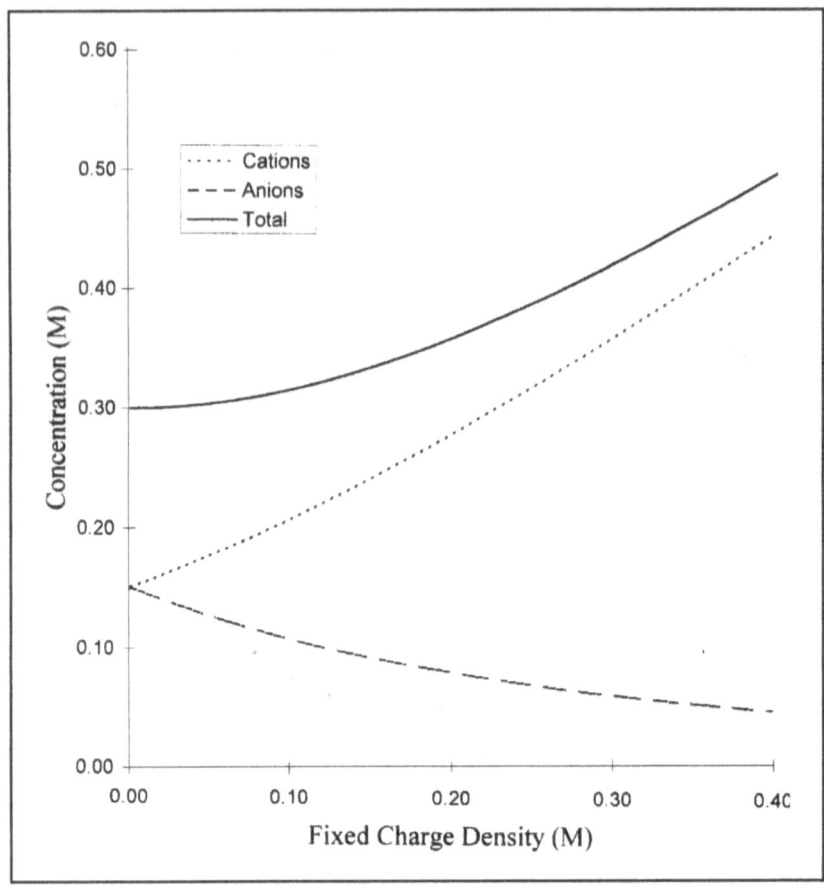

Fig. 4.2. The pericellular concentration of monovalent cations, anions, and total monovalent ions as a function of fixed negative charge density. The cation and anion concentrations are determined by multiplying the partition coefficient predicted by the Gibbs-Donnan equilibrium (see Appendix 4.1) by the monovalent cation concentration in medium outside the cartilage (external).

the total number of pericellular cations and anions, and therefore osmotic pressure, increases with fixed charged density. These relationships are derived in Appendix 4.1 and presented graphically in Figure 4.2.

The above discussion illustrates that water is the primary determinant of the mechanical properties of cartilage. The small pore size within the matrix (~2 nm)[9] severely limits the flow of water, enabling cartilage to resist compression. The high osmotic pressure of compressed cartilage aids in resisting further compression and gives the cartilage its resilience.

The effects of biochemical factors on cartilage

The synthesis, incorporation, and degradation of matrix components are influenced by both biological and mechanical factors. In vitro cartilage derived factor (CDF), which is probably insulin-like growth factor-1 (IGF-I), stimulates DNA synthesis by chondrocytes as determined by ^{3}H-thymidine incorporation and PG synthesis as determined by ^{35}S-sulfate incorporation.[11-13] IGF-1 stimulates both matrix synthesis and proliferation by articular chondrocytes, though it appears to be more potent in stimulating proliferation.[13-15] Basic fibroblast growth factor (bFGF) has a strong mitogenic affect on articular chondrocytes[15-17] and a controversial affect on matrix synthesis.[16-18] There have been conflicting results reported for the role of transforming growth factor-β (TGF-β). Some researchers have reported that TGF-β decreases the production of PG and switches collagen phenotype from type II to type I, both indicators of chondrocyte de-differentiation.[19] Others have found that TGF-β stimulates the synthesis of both collagen and GAG.[20]

These growth factors have also been reported to promote chondrogenesis in vivo. For example, bFGF promotes healing of surgically induced lesions in rabbit patellas.[21] Mechanical and biological factors may act in concert, as highlighted by the observation that interleukin-1 (IL-1) suppresses PG synthesis in normal but not immobilized joints.[22]

The effects of mechanical forces on cartilage in vivo

It has been known for more than 100 years that cartilage responds to physical forces and is able to remodel in response to the prevailing stress.[23] More recently, clinical observations and in vivo studies have revealed the extent to which this remodeling occurs. Chondrocytes of exercised rabbits are larger than those found in sedentary controls.[24] Load-bearing surfaces in joints are thicker and stronger than nearby nonload-bearing surfaces of the same joint and have higher PG concentrations.[25-27] Decreased loading or immobilization of a joint reduces PG synthesis and content, though the PG content is gradually restored once the joint is remobilized.[28-30] Studies of animals with amputated limbs demonstrate that this loss appears to result from diminished loading rather than lack of motion, since the cartilage in joints above the amputation (which would experience diminished loading) thins in the presence of continued motion of the joint. In contrast, increased loading of a joint in vivo results in stronger and thicker cartilage.[31] However, excess or impact[32] loading a normal joint or surgically modifying the joint to produce altered patterns of loading on the cartilage produces degradation of the cartilage, which serves as an animal model of osteoarthritis.[33-34] The focal lesions of osteoarthritis in the knee and hip occur together with areas of peak loading, suggesting that excessive mechanical forces are involved in its pathogenesis.[35]

The effects of mechanical forces on cartilage in vitro

In vivo models clearly show that mechanical forces influence the rates of synthesis and degradation of matrix by chondrocytes. Mechanical loading, however, produces complex changes in the electrical, chemical and mechanical environment of the cartilage, and it is difficult to know which factors are the key mediators of the response. Cell culture and ex vivo models that provide better defined electrical, chemical and mechanical environments have been instrumental in determining the mechanisms mediating the response of chondrocytes and cartilage to mechanical forces.

Static Compression

Static compression of excised cartilage ex vivo invariably results in a dose-dependent decrease in PG synthesis and collagen.[36-39] Sah et al obtained disks

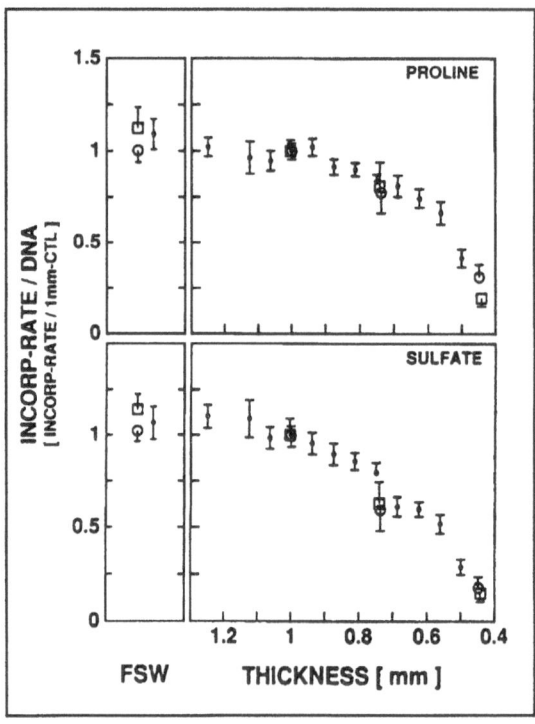

Fig. 4.3. The effect of static compression on an explanted disks of cartilage on PG and collagen synthesis as determined by ^{35}S-sulfate and ^{3}H-hydroxy-proline incorporation. Disks of bovine articular cartilage 1.0 mm thick were bathed in medium and subsequently swelled to approximately 1.25 mm (free swelling volume). Swollen disks were bathed in medium containing normal (filled circles), 2 (open squares), or 3 (open circles) times the normal concentration of proline and sulfate while compressed between two impermeable plates. There was no significant difference in the dependence of incorporation rate on the compression for disks in the medium containing 1, 2 or 3 times the normal concentration of proline and sulfate, suggesting that the compression-induced inhibition of incorporation is not due to diffusional limitations. With permission from Kim et al. Arc Biochem Biophys 1994; 311(1);1-12.

of cartilage 1 mm in thickness from a bovine femoropatellar groove and immersed them in isotonic solution. The cartilage absorbed water and swelled to a thickness of approximately 1.25 mm. The cartilage disks were then uniaxially compressed between two impermeable plates for 12 hours, decreasing the thickness to its original value, which had no significant effect on ^{35}S-sulfate incorporation, an indicator of PG synthesis, or ^{3}H-hydroxy-proline incorporation, an indicator of collagen synthesis. Further compression of these disks, however, resulted in decreased PG and collagen synthesis (Fig. 4.3).[40] Decreases in each index of matrix synthesis were first detected at 15% compression, and further compression to ~0.5 mm decreased incorporation of each tracer to approximately 25%.

Though static compression has been shown to inhibit the synthesis of two major matrix components, prolonged static compression had no effect on hyaluronan synthesis.[41] Other indications that static compression has specific effects on chondrocyte metabolism are provided by observations of cartilage recovering from static compression. After release from 50% compression for 12 hours, linked protein synthesis returned to control levels immediately while aggrecan synthesis required 60 hours.

A similar trial illustrated that static compression reduced the rate of binding of radiolabeled PG monomers to hyaluronate. Specifically, the time required for 50% of the radiolabeled PG to bind to the hyaluronate was increased; however, the maximum percentage of PG bound to hyaluronate was not affected, but merely delayed.

To understand potential mechanisms that may mediate the inhibition of extracellular matrix, it is necessary to appreciate the changes in the physical, chemical and mechanical environment induced by static compression. These changes can be classified into two groups: transient changes, which occur only while the volume of the disk is changing, and persistent changes, which occur as long as pressure is applied. A flow chart illustrating the transient and persistent changes resulting from static and dynamic loading is shown in Figure 4.4.

The initial transient change found in static compression between two parallel nonporous plates is the flow of fluid radially out of the disk. This flow of fluid causes motion among mobile ions within the disk, resulting in an electrical current. This current establishes an electrical potential, here referred to as the streaming potential. In addition, the flow of fluid relative to the chondrocytes results in the application of a tangential force (shear stress) to the cells.

The persistent changes essentially are dependent on the loss of water and subsequent increase in fixed charge density. As described above, static compression results in increased cation concentration; decreased anion

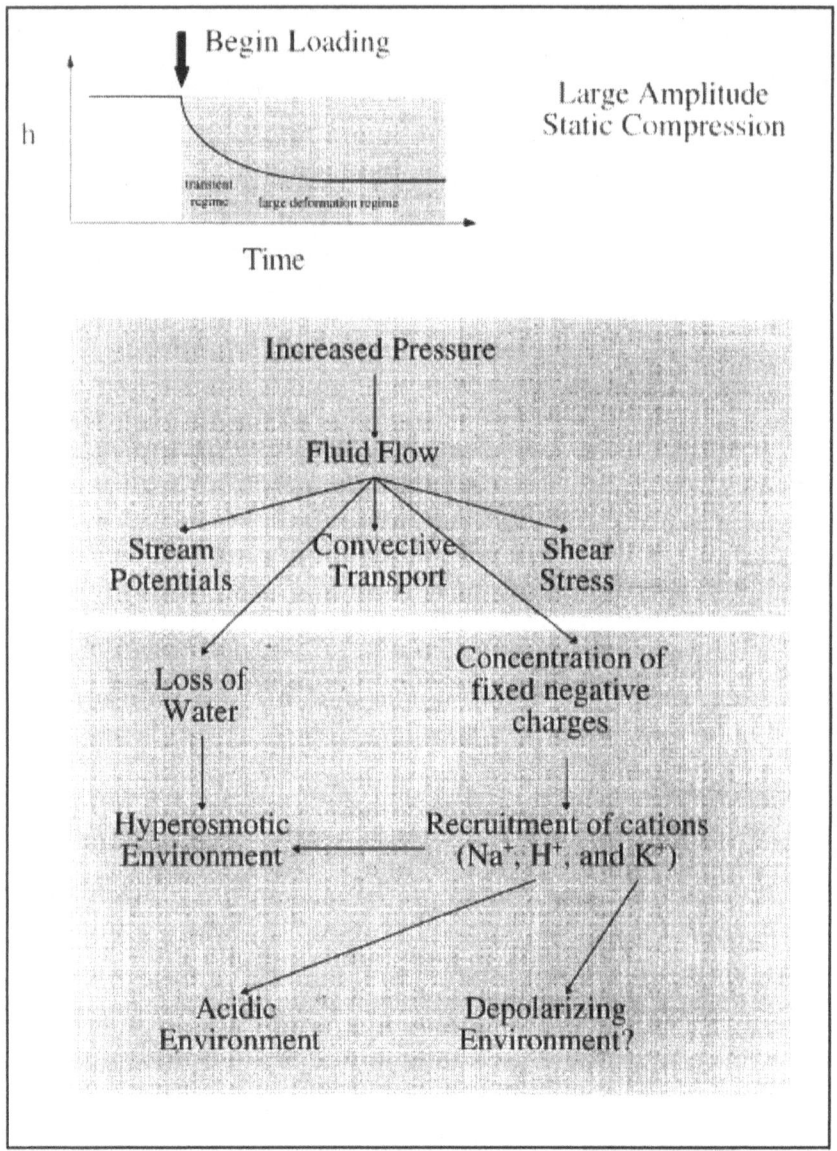

Fig. 4.4. The mechanical and electrochemical effects of static (Panel A, this page) and dynamic compression (Panel B, opposite page) of cartilage. In addition to the depicted effects, both dynamic and static compression produce cellular deformation.

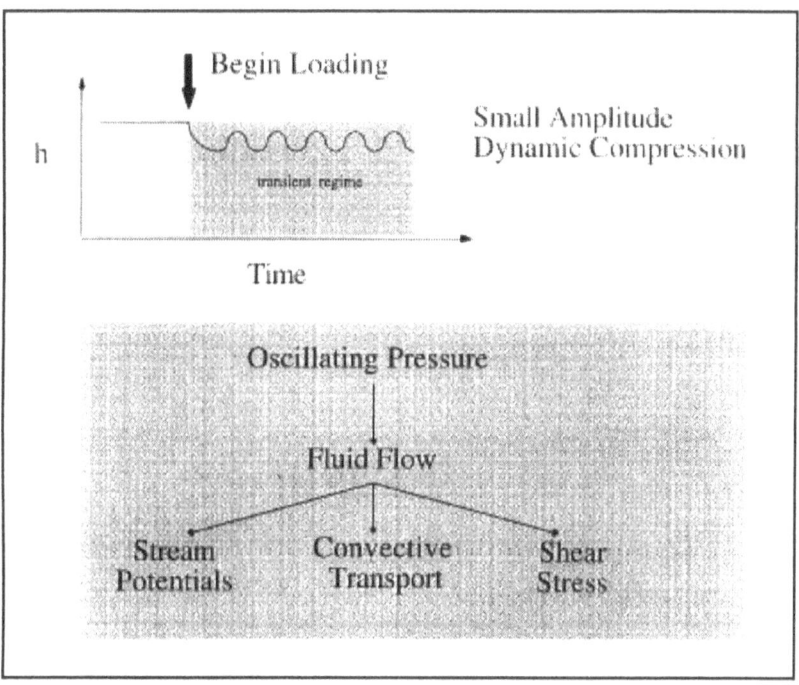

Fig. 4.4.(Panel B, see legend, opposite page).

concentration; and a hyperosmotic, acidic and possibly depolarizing environment. Any of these changes may potentially decrease the rate of PG synthesis and its binding to hyaluronate.

Researchers utilize two major tools to try to understand which of these changes are responsible for these inhibitions. The first approach involves mathematical models that seek to predict the magnitude of the parameters and determine if these predicted values correlate to experimentally observed responses. Sah et al used the Gibbs-Donnan equilibrium to predict the pericellular pH of compressed cartilage in medium with pH 7.45 and of uncompressed cartilage disks in media of varying pH. Since varying the pH of the media would modify the pericellular concentration of H^+ and OH^- but not of other ions, it is possible to isolate the influence of pH from other potential mediators. A correlation between pericellular pH and the characteristic time for ^{35}S-sulfate monomer binding (Fig. 4.5) suggests that pericellular pH directly regulates the rate of monomer binding, whereas compression and alteration of medium pH act indirectly on monomer binding rate by changing

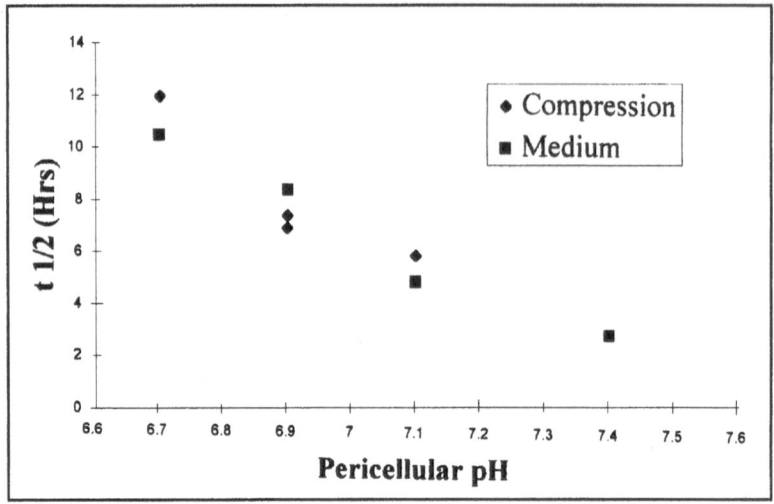

Fig. 4.5. Similar pericellular pH resulting from compression or modification of the pH of the external medium have similar effects on PG monomer binding to hyaluronate. The pericellular pH resulting from modification of external medium or compression of the cartilage disk was predicted using the Gibbs-Donnan equilibrium. Data from Biochem J 267:803-8;1990.

pericellular pH. Further support for this hypothesis is seen in an experiment where 2% oscillating loading of a disk compressed to 1.0 mm did not affect $t_{1/2}$, the characteristic time for ^{35}S-sulfate labeled monomer binding, compared to disks statically compressed to 1.0 mm. As discussed below, small-amplitude dynamic compression does not affect ion distribution in cartilage. Therefore, if pH controls response, no difference between the two types of compression would be expected.

It was suggested by Jones et al that fluid loss (which would increase Na^+ concentration) rather than load per se mediates load-induced inhibition of PG synthesis.[42] Like static compression, the addition of Na^+ or sucrose to the medium also decreases ^{35}S-sulfate incorporation. Mathematical models again are useful in interpreting this data. The Gibbs-Donnan equilibrium predicts that each of these treatments increases pericellular Na^+ concentration. Figure 4.6 illustrates that similar increases in pericellular Na^+ concentration resulting from the addition of Na^+ or sucrose to the external medium produces similar decreases in PG synthesis. Additional studies were conducted to distinguish the role of compression and high levels of pericellular Na^+. Cartilage disks were bathed in medium containing low Na^+ concentration.[43] When these

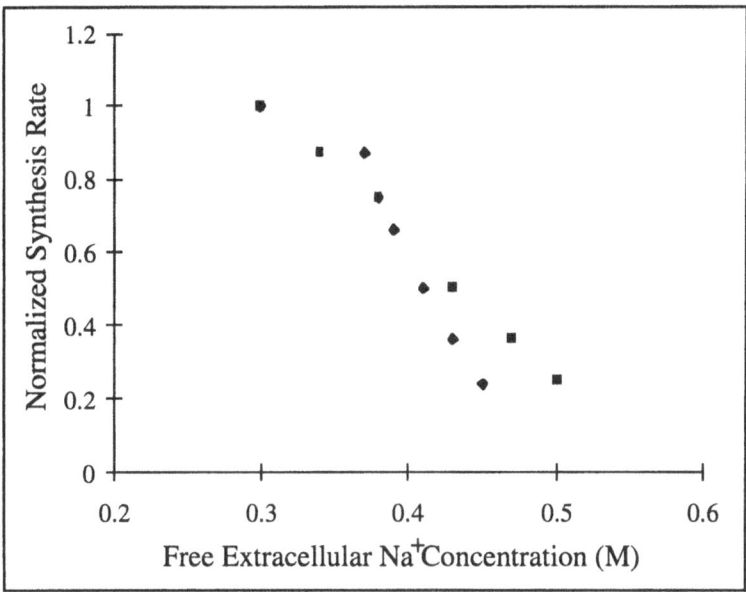

Fig. 4.6. Similar pericellular Na⁺ concentrations (x-axis) result in similar inhibition of proteoglycan synthesis (y-axis). The pericellular Na⁺ concentration resulting from addition of Na⁺ (♦) or sucrose (■) to external medium was predicted using the Gibbs-Donnan equilibrium. The similarities between Figs. 4.6 and 4.3 suggest that static compression may inhibit PG synthesis by modifying pericellular Na⁺ concentration.

disks were compressed, pericellular Na⁺ concentration increased, but did not exceed values normally found in uncompressed cartilage. PG synthesis in these disks was similar to that of uncompressed disks exposed to medium with normal Na⁺ concentration. Compression that does not result in nonphysiological pericellular Na⁺ concentration does not lead to inhibition of PG synthesis. However, since Na⁺ is the major determinant of pericellular osmotic pressure, it is difficult to determine whether the inhibition is due to elevated Na⁺ levels or elevated osmotic pressure.

The studies of chondrocytes cultured in vitro described above suggest that cell shape is not a major factor governing the metabolic response to static compression. Monolayer cultures of chondrocytes adherent to a rigid surface adopt a much flatter shape than chondrocytes in vivo. Modification of the extracellular medium would not be expected to alter cell shape beyond osmotic effects. Therefore, cells with altered initial shape that did not experience deformation mimicked the metabolic behavior of cells in cartilage explants, suggesting that cell shape is unimportant in mediating these responses. In effect, no matter how pericellular Na^+ concentrations (osmotic pressures) are increased, the result is the same: PG synthesis is inhibited. However, compression need not inhibit PG synthesis; this occurs only at high Na^+ concentrations (osmotic pressures).

Dynamic Compression

Researchers have studied the effects of small-, medium- and large-amplitude dynamic compression on PG synthesis. Results indicate that medium-amplitude dynamic compression stimulates PG synthesis. Large-amplitude dynamic compression apparently has widely varying results.[44] Small-amplitude dynamic oscillations have been more extensively studied than medium- or large-amplitude oscillations and have been especially useful in determining which physical, chemical or electrical stimuli affect chondrocyte metabolism. One of the main advantages of these experiments is that small-amplitude dynamic oscillations are similar to those occurring in vivo under most normal loading situations, such as walking. Walking results in cyclic deformation of 1 Hz with less than 1-5% deformation.[45] Larger deformations (~20%) occur in the intervertebral disks throughout the diurnal cycle. For this reason, they are very helpful in relating experimental results to actual physiological problems.

Whereas static compression results in large changes in volume and chemical environment, small-amplitude dynamic compression does not, and therefore does not affect pericellular ion levels. The initial changes resulting from small-amplitude dynamic compression are transient cell deformation and fluid flow. As in the static compression models discussed previously, fluid flow results in streaming potential, convective transport and shear stress. The key difference between the two examples is that these responses are transient in static compression models, whereas they are continuous in dynamic compression models.

An important distinction between static and dynamic models lies in the spatial distribution of fluid flow and its resulting effects. A series of experiments conducted by Kim et al indicate that stimulation of PG synthesis is exhibited even with small-amplitude dynamic compression, provided that the frequency of compression is greater than 0.001 Hz.

The frequency of applied strain apparently is of great importance. High-frequency (0.01-1 Hz) dynamic compression increased the incorporation rate of ^{35}S-sulfate and ^3H-proline into cartilage by 20-40%. Low-frequency (~0.001 Hz) dynamic compression had little effect at low amplitudes but increased the incorporation of ^{35}S-sulfate and ^{3}H-proline at medium amplitudes. Low-frequency trials run at high amplitudes yielded widely varying results.

To determine the regional effects of the applied strain, the researchers removed the central 2-mm-diameter region from the original 3-mm disk and quantified ^{35}S-sulfate incorporation (and, therefore, PG synthesis) in each region. They observed that PG synthesis was elevated in both regions at strain frequencies between 0.002 and 0.01 Hz. At 0.1 Hz, the stimulation values of the outer region was greater than that in the inner region.

Cell deformation

Chondrocytes plated on elastin membranes were subjected to 10% stretching at 1 Hz, which elicited an increase in GAG synthesis.[46] Simple agitation also elicited this response in some trials, but not in others.[47] Other studies support the assertion that cell deformation alters PG synthesis[48] and chondrocyte phenotype.[49-51] However, dynamic compression of agarose gels containing chondrocytes also stimulated PG synthesis,[52] even though these chondrocytes did not experience substantial deformation.[53,54] Taken together, these data suggest that though it may influence chondrocyte metabolism, deformation does not account for the increased PG synthesis exhibited by chondrocytes exposed to cyclic compression.

Enhanced transport

In unloaded and statically compressed cartilage, the transport of nutrients from the external fluid to the chondrocytes and the transport of metabolites in the opposite direction is achieved by diffusion. Cyclic loading and the resulting continuous flow of fluid in and out of the cartilage increases the exchange of some compounds between the cartilage and the surrounding fluid. In an experiment designed to approximate the cyclic loading that occurs in walking (2.8 MPa at 1 Hz), O'Hara et al observed that the absorption of small molecules (radiolabeled urea and sodium iodide) by human femoral head cartilage was not noticeably enhanced by fluid transport.[55] A similar study conducted by Maroudas et al revealed no difference in transport rates of small dye molecules in the presence or absence of cyclic loading of cartilage.[56] These experiments indicate that the rate of diffusion for small molecules greatly supersedes any effect caused by fluid flow.

The transport rate of some large solutes, however, is greatly affected by cyclic loading. Under the same conditions described above, O'Hara et al observed that the desorption rate of human serum albumin from cartilage increased from 30 to 100% under cyclic loading conditions. These results suggest that the movement of macromolecules such as hormones, enzymes and growth factors into and out of cartilage is significantly affected by cyclic loading.

Fluid flow resulting from cyclic loading of cartilage affects absorption and desorption rates of large molecules; this effect diminishes as the size of the molecules decreases. In the case of most nutritional molecules needed by cartilage (such as glucose and oxygen), the effect of fluid flow on absorption/ desorption rate is negligible but can affect uptake of macromolecules such as growth factors that are important in cell metabolism. The effects of fluid flow as convective transporter of molecules is fairly well understood. An interesting question raised by Kim et al[57] is whether fluid flow elicits changes solely by convection or whether it also directly stimulates chondrocytes by the shear stress and streaming potentials it generates.

Flow

Researchers have taken advantage of the regional variations in flow and pressure by comparing the metabolic responses of different regions of compressed cartilage. In mathematical models of dynamic compression, the radial velocity of fluid flow increases with distance from the center of the disk. In contrast, pressure decreases from a maximum value in the center of the disk to a minimum (approaching atmospheric pressure) at the radial edge. These relationships are derived for slow compression without radial expansion in Appendix 4.2 and are graphically illustrated in Figure 4.7. The derivation for compression resulting in radial deformation of the cartilage is more involved.[58] The velocity and pressure profiles expected in compressed cartilage with or without radial expansion follow the same general trends, and approximate values are shown in Figure 4.7. By comparing predicted values of velocity and pressure with regional PG synthesis, Kim et al[57] concluded that PG synthesis co-localized with regions of high fluid velocity but was inversely related to pressure. Subsequently, Smith and coworkers demonstrated that monolayers of bovine and human articular chondrocytes exposed to laminar fluid flow increased GAG synthesis, prostaglandin E_2 release, and mRNA levels for tissue inhibitors of metalloproteinase compared to stationary controls.[59] The maximum pressure calculated for the dynamic compression was 0.5 MPa, significantly less than the minimum pressure reported to stimulate PG synthesis by bovine chondrocytes (see below).

Pressure

Other researchers have conducted experiments to investigate the influence of pressure on PG synthesis.[60-62] These researchers used a pressurized chamber rather than uniaxial compression plates to increase hydrostatic pressure. This method of applying pressure is uniform over the surface of the disk and does not result in volume changes or fluid flow. Hall et al found that

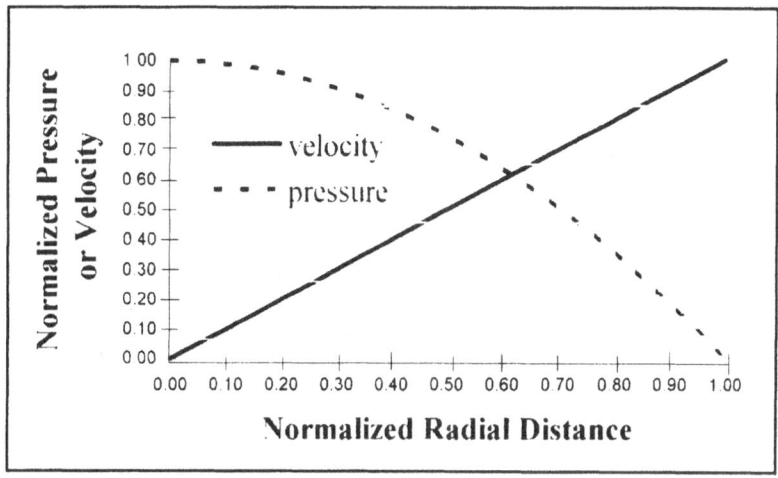

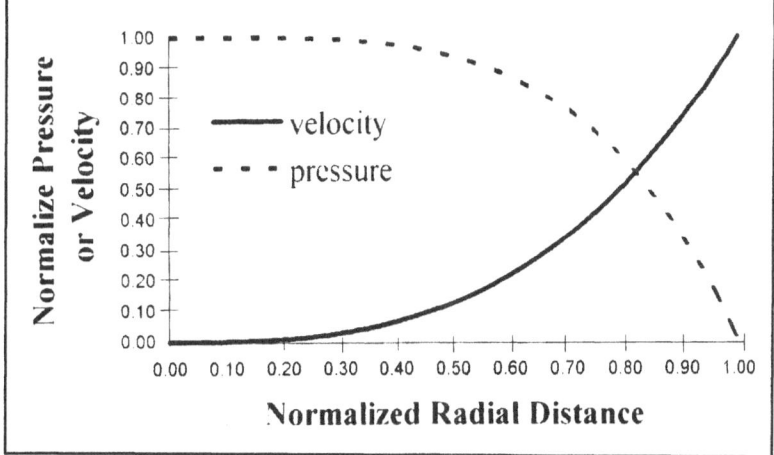

Fig. 4.7. Velocity and pressure profile in compressed cartilage as a result of slow (Panel A, top) and rapid (Panel B, bottom) compression. Velocity and pressure profiles for slow compression are calculated from equations derived in Appendix 4.2; the profiles shown in Panel B are approximations based on figures from Kim et al.[57]

application of hydrostatic pressures <10 MPa for 2 hours resulted in PG synthesis stimulation, whereas higher pressures (30-50 MPa) applied for the same duration inhibited PG stimulation. However, if the pressure was applied for only 20 s, pressures from 10-20 MPa stimulated PG synthesis while applied pressure of 50 MPa had no effect. These data show that both the magnitude and duration of the applied pressure are important in determining the metabolic response of excised cartilage.

Different types of cartilage tend to respond to the pressure they normally are subjected to in the body. For instance, bovine cartilage synthesis rates were increased when subjected to 5-15 MPa of pressure for 20 s, which is comparable to its normal physiological load. Intervertebral disc showed a decrease in synthesis when subjected to the same conditions, but showed increased synthesis when subjected to a load of 1.0-2.5 MPa, which approximates its load in vivo.[63]

Summary

Cartilage is exposed to large-magnitude (0.1-20 MPa) mechanical forces in vivo. The unique composite proteoglycan gel/collagen mesh composition of cartilage enables it to withstand large compressive loads, primarily by retaining water. The presence of fixed negative charges associated with proteoglycans in cartilage results in a higher concentration of mobile cations and a lower concentration of mobile anions, leading to a hyperosmotic, acidic environment in the extracellular space of cartilage.

In vivo cartilage remodels in response to mechanical forces; cartilage on load-bearing surfaces in joints is thicker and stronger than that of adjacent non-load-bearing surfaces. Decreased loading of a joint results in diminished proteoglycan content and cartilage thickness. Static compression in vitro results in decreased production of GAG and collagen, probably by increasing pericellular ionic concentration. Small-amplitude cyclic compression stimulates both GAG and collagen synthesis. Mathematical models suggest that stimulation of matrix synthesis is associated with fluid flow rather than increased pressure. Fluid flow resulting from cyclic loading of cartilage facilitates the transport of materials within the cartilage. The fluid flow may act by increasing nutrient transport, or it may affect synthesis by generating streaming potential or shear stress.

Appendix 4.1. Derivation of the partition coefficient, k, for single charged cations or anions as a function of fixed charged density.

C^+ = Concentration of cations outside cartilage

C^- = Concentration of anions outside cartilage

$\overline{C}^+$ = Concentration of cations inside cartilage

$\overline{C}^-$ = Concentration of anions inside cartilage

$$\frac{\overline{C}^+}{C^+} = k \qquad (2) \quad \text{Gibbs-Donnan equilibrium for anions}$$

$$\frac{C^-}{\overline{C}^-} = k \qquad (3) \quad \text{Charge neutrality inside cartilage}$$

$$\overline{C}^+ = \overline{C}^- + FCD \qquad (4) \quad \text{Charge neutrality outside cartilage}$$

$$C^+ = C^- \qquad (5) \quad \text{Equations 1 \& 2 substituted into 3}$$

$$kC^+ = \frac{C^-}{k} + FCD \qquad (6) \quad \text{Equation 4 substituted into 5}$$

$$kC^+ = \frac{C^+}{k} + FCD$$

$$k = \frac{FCD + \sqrt{FCD^2 + 4(C^+)^2}}{2C^+} \qquad (7) \quad \text{Solving equation 6 for k}$$

h   =   height of disk
R   =   radius of disk
r   =   radial distance from center of the disk
v   =   volume of a disk of radius r
u_r =   average velocity of fluid at distance r from the center of disk
P_r =   pressure at distance r from the center of disk
P_r =   pressure at the radially outer edge of the cartilage
$\dfrac{dh}{dt}$ =   displacement as a function of time

$$\frac{dv}{dt} = \frac{dh}{dt} \cdot \pi r^2 \qquad \text{change in volume of a disk of radius r resulting from displacement} \qquad (1)$$

$$A = 2\pi r \cdot h \qquad \text{cross sectional area through which fluid passes} \qquad (2)$$

$$u_r = \frac{1}{a} \cdot \frac{dv}{dt} \qquad (3)$$

Substituting equations 1 and 2 into 3 gives,

$$u_r = \frac{2}{h}\frac{dh}{dt}r \qquad (4)$$

The pressure gradients and velocity are interrelated by the equation

$$u_r = -K_H \frac{dp}{dr} \quad \text{where } K_H \text{ a proportionality constant defined as hydraulic conductivity} \qquad (5)$$

Equating equations 4 and 5 yields

$$\frac{2}{h}\frac{dh}{dt}r = -K_H \frac{dp}{dr} \qquad (6)$$

Separating the variables and applying limits of integration yields

$$\frac{2}{h}\frac{dh}{dt}\int_R^r r\,dr = -K_H \int_R^r dp \qquad (7)$$

which upon integrating gives

$$\frac{2}{h}\frac{dh}{dt}(r^2 - R^2) = -K_H \cdot (P_r - P_R) \qquad (8)$$

Recognizing that P_R or the pressure on the outer surface of the cartilage equals
the pressure of fluid outside of cartilage which is approximately 0 and rearranging yields,

$$P_r = \frac{2}{hK_H}\frac{dh}{dt}(R^2 - r^2). \qquad (9)$$

Appendix 4.2. Derivation of the velocity and pressure profiles within a disk of cartilage slowly compressed between two impermeable plates.

References

1. Mizrahi J, Solomon L, Kaufman B et al. J Bone Joint Surg 1981; 63B:610-13.
2. Tepic S, Macirowski T, Mann RW. Proc Summer Comp Simulation Conf 1984; 2:834-9.
3. Hodge WA, Fijan RS, Carlson KL et al. Contact pressures in the human hip joint measured in vivo. Proc Natl Acad Sci USA 1986; 83:2879-83.
4. Stockwell RA. The interrelationship of cell density and cartilage thickness in mammalian articular cartilage. J Anatomy 1971; 109:411-421.
5. Mayne R, von der Mark K. Collagens of cartilage. In: Cartilage, Structure, Function and Biochemistry. Vol. 1. New York: Academic Press, 1983:181-214.
6. Byers R, Bayliss MT, Maroudas A et al. Hypothesizing about joints. In: Maroudas A, Holborrow J, eds. Studies in Joint Disease. London: Pitman Medical, 1983:241-76.
7. Voet D, Voet JG. Sugars and polysaccharides. In: Biochemistry. New York: John Wiley & Sons, 1990:260.
8. Comper WD. Physicochemical aspects of cartilage extracellular matrix. In: Hall B, Newman S, eds. Cartilage: Molecular Aspects. Boca Raton: CRC Press, 1991:73.
9. Maroudas A, Mizrahi J, Katz EP et al. Physicochemical properties and functional behavior of normal and osteoarthritic human cartilage. In: Kuettner K, ed. Articular Cartilage Biochemistry. New York: Raven, 1986:311-27.
10. Kato Y, Nomura Y, Daikuhara N et al. Cartilage-derived factor (CDF). I. Stimulation of proteoglycan synthesis in rat and rabbit costal chondrocytes in culture. Exp Cell Res 1980; 130:73-81.
11. Kato Y, Nomura Y, Tsuji H et al. Cartilage-derived factor (CDF). II. Somatomedin-like action on cultured chondrocytes. Exp Cell Res 1981; 132:339-47.
12. Seyedin SM, Rosen DM. Cartilage growth and differentiation factors. In: Hall B, Newman S, eds. Cartilage: Molecular Aspects. Boca Raton: CRC Press, 1991:135.
13. Trippel SB. Growth factor actions on articular cartilage. J Rheum 1995; 22(1) suppl 43:129-32.
14. Luyton FP, Hascall CV, Nissley SP et al. Insulin-like growth factor maintained steady state metabolism of proteoglycans in bovine articular cartilage explants. Arch Biochem Biophys 1988; 267:416-25.
15. Osborn KD, Trippel SB, Mankin HJ. Growth factor stimulation of adult articular cartilage. J Orthop Res 1989; 7:35-42.
16. Jones KL, Addison J. Pituitary growth factor as a stimulator of growth in culture of rabbit articular chondrocytes. Endocrinology 1975; 97:359-65.
17. Sachs BL, Goldberg VM, Moskowitz RW et al. Response of articular chondrocytes to pituitary fibroblast growth factor (FGF). J Cell Physiol 1982; 112:51-9.

18. Kato Y, Gospodarowicz D. Sulfated proteoglycan synthesis by confluent cultures of rabbit costal chondrocytes grown in the presence of fibroblast growth factor. J Cell Biol 1984; 100:477-85.

19. Rosen DM, Stempien SA, Thompson AY et al. Transforming growth factor-beta modulates the expression of osteoblast and chondroblast phenotypes in vitro. J Cell Physiol 1988; 134:337-46.

20. Redini F, Lafuma C, Pujol J-P et al. Effects of cytokines and growth factors on the expression of elastase activity by hyman synoviocytes, dermal fibroblasts, and rabbit articular chondrocytes. Biochem Biophys Res Commun 1988; 155:786-93.

21. Seyedin SM, Rosen DM. Cartilage growth and differentiation factors. In: Hall B, Newman S, eds. Cartilage: Molecular Aspects. Boca Raton: CRC Press, 1991:139.

22. van Lent PLEM, van de Loo FAJ, van den Bersselaar et al. Chondrocyte nonresponsiveness of arthritic articular cartilage caused by short-term immobilization. J Rheumatol 1991; 18:709-15.

23. Helminen H, Jurvelin J, Kiviranta I et al. Joint loading effectson articular cartilage: a historical review. In: Helminen H, Kiviranta I, Tammi M, eds. Joint Loading: Biology and Health of Articular Structures. Bristol: John Wright, 1987:1-46.

24. Paukkonen K, Selkainaho K, Jurvelin J et al. Cells and nuclei of articular cartilage chondrocytes in young rabbits enlarged after nonstrenuous physical exercise. J Anat 1985; 142:13-20.

25. Bjelle AO. Content and composition of glycosaminoglycans in human knee joint cartilage. Variation with site and age in adults. Conn Tiss Res 1975; 3:141-7.

26. Roberts S, Weightman B, Urban JPG et al. Mechanical and biochemical properties of human articular cartilage in osteoarthritic femoral heads and in autopsy specimens. J Bone Joint Surg [Br]1986; 68:278-88.

27. Slowman SD, Brandt KD. Composition and glycosaminoglycan metabolism of articular cartilage from habitually loaded and habitually unloaded sites. Arthritis Rheum 1986; 29:88-94.

28. Jurvelin J, Helminen H, Lauritsalo S. Influences of joint immobilization and running exercise on articular cartilage surfaces of young rabbits. Acta Anat 1985; 122:62-8.

29. Palmoski M, Perricone E, Brandt KD. Development and reversal of a proteoglycan aggregation defect in normal canine knee cartilage after immobilization. Arthritis Rheum 1979; 22:508-17.

30. Saamenen AM, Tammin M, Kiviranta I et al. Maturation of proteoglycan matrix in articular cartilage under increased and decreased joint loading. Conn Tiss Res 1987; 16:163-75.

31. Kiviranta I, Tammi M, Jurvelin J et al. Moderate running exercise augments glycosaminoglycans and thickness of articular cartilage in the knee joint of young beagle dogs. J Orthop Res 1988; 6:188-95.

32. Afoke NYP, Byers PD, Hutton WC. Contact pressures in the human hip joint. J Bone Joint Surg 1987; 69B:536-42.

33. Muir H, Carney SL. Pathological and biochemical changes in cartilage and other tissues of the canine knee rsulting from induced joint instability. In: Helminen HJ, Kiviranta I, Tammi M, eds. Joint Loading: Biology and Health of Articular Structures. Bristol: John Wright, 1987:47-63.

34. Hoch DH, Grodzinsky AJ, Koob TJ et al. Early changes in material properties of rabbit articular cartilage after menisectomy. J Orthop Res 1983; 1:4-12.

35. Dieppe P, Kirwan J. The localization of osteoarthritis. Br J Rheumatol 1994; 33:201-4.

36. Gray ML, Pizzanelli AM, Lee RC et al. Kinetics of the chondrocyte biosynthetic response to compressive loading and release. Biochim Biophys Acta 1989; 991:415-25.

37. Sah RL, Grodzinsky AJ, Plaas AHK et al. Effects of tissue compression on the hyaluronate binding properties of newly synthesized proteoglycans in cartilage explants. Biochem J 1990; 267:803-8.

38. Sah RL, Doong JY, Grodzinsky AJ et al. Effects of compression on the loss of newly synthesized proteoglycans and proteins from cartilage explants. Arch Biochem Biophys 1991; 286:20-9.

39. Larsson T, Aspden RM, Heinegard D. Effects of mechanical load on cartilage matrix biosynthesis in vitro. Matrix 1991; 11:388-94.

40. Sah RL, Kim YJ, Doong JH et al. Biosynthetic response of cartilage explants to dynamic compression. J Orthop Res 1989; 7:619-36.

41. Kim YJ, Grodzinsky AJ, Plaas AHK et al. The differential aspects of static compression on the synthesis of specific cartilage matrix components. Trans Am Orthop Res Soc 1992; 17:108.

42. Jones IL, Klamfeldt A, Sandstrom T. The effect of continuous mechanical pressure upon the turnover of articular cartilage proteoglycans in vitro. Clin Orthop 1982; 165:283-9.

43. Schneiderman R, Kevet D, Maroudas A. Effects of mechanical and osmotic pressure on the rate of glycosaminoglycogen synthesis in the human adult femoral head cartilage: an in vitro study. J Orthop Res 1986; 4:393-408.

44. Parkkinen JJ, Lammi MJ, Helminen HJ et al. Local stimulation of proteoglycan synthesis in articular cartilage explants by dynamic compression in vitro. J Orthop Res 1992; 10:610-20.

45. Weightman B, Kempson G. Load carriage. In: Freeman MAR, ed. Adult Articular Cartilage. London: Pitman Medical, 1979; 293-341.

46. Lee RC, Rich JB, Kelley KM. A comparison of in vitro cellular responses to mechanical electrical stimulation. Am Surg 1982; 48:567-74.

47. Armstrong CG, Lai WM, Mow VC. J Biomech Eng 1984; 106:165-73.

48. Watt FM. The extracellularx and cell shape. Trends Biochem Sci 1986;11:482-5.

49. Benya PD, Brown PD, Padilla SR. Microfilament modification by dihydrocytochalasin B causes retinoic acid-modulated chondroctyes to

reexpress the differentiated collagen phenotype without a change in shape. J Cell Biol 1988;106:161-70.

50. Benya PD, Shaffer JD. Dedifferentated chondrocytes reexpress the differentiated collagen phenotype when cultured in agarose gels.Cell 1982;30:215-24.

51. Brown PD, Benya PD. Alterations in chondrocyte cytoskeletal architecture during phenotypic modulation by retinoic acid an dihydrocytochalasin B-induced reexpression. J Cell Biol 1988;106:171-9.

52. Buschmann MD, Gluzband YA, Grodzinsky AJ. Mechanical compression modulates matrix biosynthesis in chondrocyte / agarose culture. J Cell Sci 1995;108:1497-1508.

53. Freeman PM, Natarajan R, Kimura JH. Chondrocyte cells respond mechanically to compressive loads. J Orthop Res 1994 12:311-20.

54. Lee DA, Bader DL Alterations in chondrocyte shape in agarose in response to mechanical loading and its response to matrix production. Trans Orth Res Soc 1994;19:103.

55. Ohara BP, Urban JPG, Maroudas A. Influencce of cyclic loading on the nutrition of articular cartilage. [Ann Rheumatic Diseases] 1990;49:536-9.

56. Maroudas A, Bullough PG, Swanson SA et al. The permeability of articular cartilage. J Bone Joint Surg [Br] 1968;50:166-77.

57. Kim Y, Sah RLY, Grodzinsky AJ. Mechanical regulation of cartilage biosynthetic behavior: physical stimuli. 1994;311(1):1-12.

58. Spilker RL, Suh JK, Mow VC Effects of friction on the unconfined compressive response of articular cartilage: a finite element analysis. 1990;112:138-46.

59. Smith RL, Donlon BS, Gupta MK et al. Effects of fluid-induced shear on articular chondrocyte morphology and metabolism in vitro. J Orthop Res 1995;13(6):824-31.

60. Hall AC, Urban JPG, Gehl KA. The effects of hydrostatic pressure on matrix synthesis in articular cartilage. J Orthop Res 1991; 9:1-10.

61. Kimura JH, Schipplein OD, Kuettner KE et al. Effects of hydrostatic loading on extracellular matrix formation. Trans Orthop Res Soc 1985; 10:365.

62. Lippiello L, Kaye C, Neumata T et al. In vitro metabolic response of articular cartilage segments to low levels of hydrostatic pressure. Connect Tissue Res 1985; 13:99-107.

63. Ishihara H, Urban JPG, Hall AS. The effect of physiological hydrostatic pressures on synthesis in different regions of the invertebral disc. J Physiol [in press].

CHAPTER 5

Muscle Cells

Muscle Structure and Function

Perhaps the most readily noticeable example of a biological response to physical forces is the hypertrophy of skeletal muscle. The prominence of this effect probably is derived from the large percentage of body mass (~40%) composed of skeletal muscle. Cardiac and smooth muscle also respond dramatically to mechanical loading, but these responses are not as evident, probably due to their anatomical location and size (~10% of body mass).

Skeletal muscle is composed of fibers extending the entire length of the muscle. These fibers, which are between 10 and 80 microns in diameter, are made up of terminally differentiated myotubes. Based on its ability to generate forces, skeletal muscle can be further subclassified into fast-twitch and slow-twitch muscle.

The three major types of cardiac muscle (atrial, ventricular, and excitatory/conductive fibers) are composed of terminally differentiated mononuclear myocytes. The contraction of atrial and ventricular fibers resembles that of skeletal muscle except that the length of the contraction is much greater. The excitatory/conductive fibers contract very weakly; instead of generating mechanical forces, they initiate and propagate the electrical stimuli that control the contraction of other fibers.

Smooth muscle fibers are much smaller than skeletal muscle fibers, having a diameter of 2-5 microns and lengths ranging from 20 to 500 microns. Smooth muscle can be subclassified into multiunit, or visceral, smooth muscle and single-unit smooth muscle. Multiunit smooth muscle is composed of discrete muscle fibers, each acting independently. Examples of these muscles can be found in the ciliary and iris of the eye. Single-unit smooth muscle is a mass of individual fibers (numbering from hundreds to millions of fibers) that is physically and electrochemically (via gap junctions) interconnected and acts in concert to form a functional syncytia.

Mechanical Forces: Their Effects on Cells and Tissues, by Keith J. Gooch and
Christopher J. Tennant. © 1997 Landes Bioscience.

The effects of mechanical forces on muscle in vivo

Skeletal Muscle

Skeletal muscles have a remarkable ability to remodel to match the mechanical demands placed upon them. Although some forms of exercise, such as constant stimulation at 10 Hz, have been shown to cause muscle atrophy,[1] muscle usually responds to exercise by increasing its mass, especially if the muscle is stretched during the contractile process. If a muscle is stretched beyond its normal length, additional sarcomeres are added to the fiber, increasing its length. Conversely, if a muscle remains shortened for a prolonged period of time, the length of the fibers decreases. By lengthening or shortening in response to stretch, the muscle can maintain a length appropriate for its environment. This growth is almost exclusively the result of an increase in the size of individual fibers (fiber hypertrophy) rather than cellular proliferation (hyperplasia). Though the precise mechanism of hypertrophy is not known, it likely results from the increased synthesis and concurrent decrease in degradation of muscle.

In addition to modifying size in response to mechanical load, muscle alters its gene expression to increase or decrease concentrations of metabolic and contractile proteins. A notable example of this biochemical remodeling occurs in the slow-twitch muscle of marathon runners, which contains increased concentrations of myoglobin, mitochondria, oxidative enzymes and capillary density compared to unexercised controls. Conversely, slow-twitch soleus (hindlimb) muscle in adult rats put into a non-weight-bearing environment for an extended time suffered a 55% loss of muscle mass and a major decline in the amount of myofibrillar protein present. After 24 days under these conditions, myofibrillar protein content was just 20% that of control animals.[2] Using a mathematical model, Booth and Kirby subdivided this progression into three phases to more clearly illustrate the points at which gene expression changes occur (Fig. 5.1).

In the first phase, which lasted only a few days, a dramatic decline in the synthesis of myofibrillar protein was accompanied by a significant decline in protein degradation.[3] Levels of α-actin mRNA and β-myosin heavy chain mRNA were unchanged, suggesting that reduced translation was responsible for the decline in protein synthesis.[2]

In the second phase, which began ~3 days after initiation, protein degradation began to increase while synthesis continued to decline.[3,4] A similar study reported this increase occurring after 2 days.[5] Myofibrillar protein levels begin to drop precipitously in this phase. This reduced level of protein synthesis may be attributable to alterations in pretranslational control, or may have been caused by a combination of pretranslational and translational changes;[3] although α-actin mRNA levels have been shown to decrease after

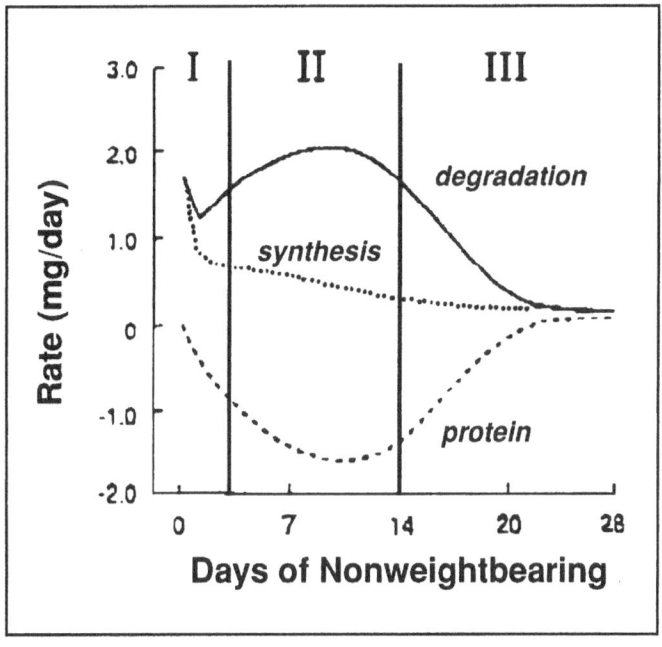

Fig. 5.1. Three phases of skeletal muscle atrophy resulting from nonweight bearing as defined by rates of protein synthesis and degradation. The difference between these rates determines the rate of protein loss. As a result of nonweight bearing, both synthesis and degradation rates decrease, though the effect on degradation rate is greater and results in net protein loss. After about 4 weeks, a new balance between synthesis and degradation is reached. With permission from Booth FW and Kirby CR. Changes in skeletal muscle gene expression consequent to altered weight bearing. Am J Physiol 1992; 262:R329-32.

7 days, β-myosin heavy chain mRNA levels remained constant over this same period.[4]

The third phase is marked by a slowing of myofibrillar protein degradation, which eventually (~24 days) reaches a steady state with protein synthesis.[3] At this point, myofibrillar protein content was measured to be just 20% of control levels.

Booth and Kirby concluded that non-weight-bearing trials with adult rats initially affect the control of mRNA translation in slow-twitch muscle, although similar experiments with juvenile rats resulted in early pretranslational control as well. Longer-term trials (56 days) resulted in a decrease of myosin heavy chain proteins from 13.4 mg to 2.1 mg per pair of soleus muscles.[2] These conditions are believed to affect pretranslational, translational, and post-translational control mechanisms.[3]

Sporadic exercise appears to increase protein synthesis and muscle mass beyond the exercise period. It is theorized that both proteolysis and synthesis increase with this type of stimulus. A 10-week training course of twice-weekly sporadic exercise resulted in a 34-67% increase in the amount of skeletal α-actin mRNA in rat tibialis anterior,[6] suggesting that upregulation of pretranslational control occurred following the exercise. But no changes in mRNA level were determined after a single bout of sporadic stimulation, which suggests that translational changes occurred to enhance protein synthesis in the short term. Extended non-weight-bearing trials conducted on the hindlimb of adult rats had little effect on fast-twitch muscle; muscle atrophy did not approach that of slow-twitch muscle, and fast-twitch myosin heavy chain protein increased 100%.[3]

Another noteworthy property of muscle fibers is their ability to convert their contractile proteins from one isoform to another to maximize their efficiency,[7] which ultimately results in a conversion of fast-twitch muscle tissue to slow-twitch tissue. For instance, fast-twitch rabbit tibialis muscle tissue converted to slow-twitch tissue in the presence of repetitive low-frequency stimulation.[8] Conversion of this type is accomplished via protein isoform changes in both myosin heavy and light chains[7] as well as in regulatory proteins within the thin filament. This conversion is carefully ordered and is believed to be dependent on altered gene expression and protein synthesis.[7]

Cardiac Muscle

Chronic hypertension (hemodynamic overload) leading to myocardial hypertrophy is of substantial clinical significance and has been widely reviewed.[9,10] Though myocardial hypertrophy is a physiological adaptation that may be observed in healthy individuals such as conditioned athletes, valvular disease or inadequate perfusion of the hypertrophied heart may contribute to heart failure, a common cause of death.

In addition to increased cell volume, hemodynamic overload results in stimulation of second-messenger pathways and altered expression of specific genes by myocytes. These genes can be classified into two groups, immediate early genes and late-response genes, according to the time course of their expression. Pressure overload in vivo induces many immediate early genes such as c-*fos*, c-*jun*, and c-*myc*, as well as heat shock protein 70 (HSP70). It appears that mechanical forces in vivo activate specific signal-transduction pathways, which in turn activate immediate early gene expression. Immediate early genes regulate late gene expression, which in turn controls the adaptive response of myocytes to physical forces.

Another effect of hemodynamic overload is isoform changes in myosin heavy chains. For example, overload of ventricular tissue is accompanied by induction of β-myosin heavy chain (β-MHC). Rats with ventricular myosin levels of 0-10% β-MHC that were exposed to hemodynamic overload experienced increases in β-MHC levels to ~80% of total myosin, which corresponded to the level of hypertrophy.[11,12] However, hypertrophy in human and pig ventricular muscle did not significantly affect myosin isoform ratios except for a loss of a small amount of atrial-type alpha-alpha homodimer MHCs.[13]

Hemodynamic overload of atrial muscle induces a change from atrial (α-MHC) to ventricular (β-MHC) myosin isoforms. The ventricular isoform hydrolyzes ATP more slowly, so contraction is slower than in the atrial form but the development of tension is more efficient.[13]

Smooth Muscle

The vascular system has the ability to respond both acutely and chronically to changes in the mechanical environment. Though visceral muscle also responds and adapts to mechanical forces, vascular smooth muscle will be examined here as a well-studied system to demonstrate principles applicable to either muscle type.

Acute effects of mechanical forces on vascular smooth muscle in vivo

The myogenic response, contraction of a blood vessel resulting from an increase in blood pressure, was originally described by Bayliss at the turn of the century.[14] Conversely, the reduction of blood pressure below normal values results in vasodilation. In each case, the response to altered blood pressure begins in seconds, and steady state is reached within minutes. Although the physiological significance of the myogenic response has yet to be unequivocally defined, it likely plays an important role in establishing basal vascular resistance and controlling blood flow and capillary pressure.

The myogenic response may be important in regions such as the brain where it is essential to maintain constant rates of perfusion in the face of varying blood pressure. An increase in systemic blood pressure results in an increased force driving blood through a vessel. This increased driving force is offset by increased vascular resistance due to vessel contraction, which tends toward maintaining blood flow at a nearly constant value. Conversely, a decrease in blood pressure may result in vasodilation, thereby decreasing vascular resistance to flow (see Fig. 5.2 for a more detailed explanation of how the myogenic response may regulate local blood flow).

In addition to controlling local blood flow, the myogenic response may help regulate capillary pressure. As an important determinant of the rate and direction of filtration across capillary beds, capillary pressure must be maintained within a specific range. Increases in systemic blood pressure may be partially offset by increased vascular resistance, resulting in relatively stable capillary pressure (see Fig. 5.3 for a more detailed explanation of how the myogenic response may regulate capillary pressure).

As its name implies, the origin of the myogenic response lies in vascular smooth muscle. Although vascular smooth muscle cells ultimately generate the force leading to vessel contraction, it has been suggested that the endothelium is the actual sensor of pressure.[15,16] The endothelium may release factors regulating the contraction of the underlying smooth muscle cells. However, experimental data[17-22] reviewed by Meininger[23] demonstrate rather convincingly that the myogenic response is endothelium-independent.

Based on the experimental results reviewed by Meininger, it is assumed that the stimulus regulating the myogenic response is detected by a component of the vascular wall, probably by smooth muscle cells. The identity of the stimulus, however, is not obvious; four potential scenarios are discussed below.

1. Smooth muscle cells may directly sense changes in pressure. While there is experimental evidence of vascular smooth muscle cells responding to pressure in vitro, it is unlikely that this is the stimulus perceived in vivo. The pressure changes along the vascular tree, with high pressures in the large arteries and a progressive decrease as blood moves to smaller arteries, arterioles, and capillaries. In addition, a radial pressure gradient exists across the vessel wall, with the highest pressure near the lumen. Both of these pressure gradients would make it very difficult for the vessel to respond in a coordinated manner to pressure, as smooth muscle cells in each region would need to have different set point pressures. More likely, the stimulus is something that is fairly constant regardless of location. Think of local regulation of blood flow: total blood flow is many orders of magnitude greater in larger arteries than small arterioles. Instead of responding to a stimulus that varies so greatly, the vessel responds to wall shear stress, a relatively constant force throughout the vasculature that is related to both blood flow and vessel diameter.

2. Wall tension is one parameter that is related to blood pressure but varies less throughout the vasculature regardless of position in the vessel wall. The greater pressures of the larger vessels are supported by thicker walls. Modeling the vessel as a cylinder in which the thickness of the vessel wall is much smaller than the vessel radius, the tension in the vessel wall, s, is governed by the equation $s = \dfrac{rP}{t}$ where r is the vessel radius and P is the pressure drop across a vessel wall of thickness t. Assuming wall thickness does not change, if wall

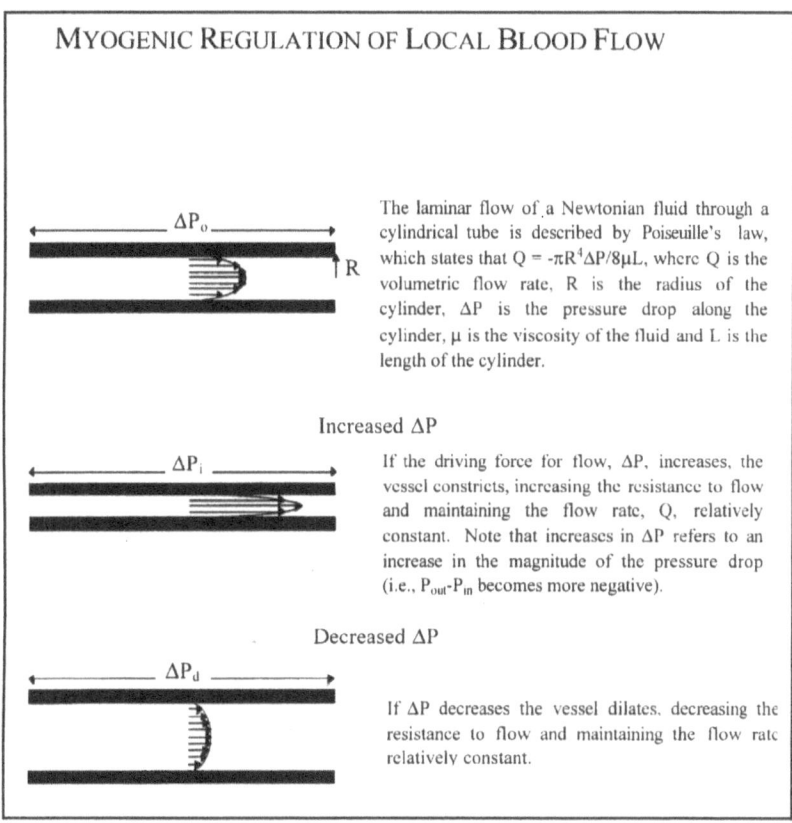

Myogenic Regulation of Local Blood Flow

ΔP_o

R

The laminar flow of a Newtonian fluid through a cylindrical tube is described by Poiseuille's law, which states that $Q = -\pi R^4 \Delta P/8\mu L$, where Q is the volumetric flow rate, R is the radius of the cylinder, ΔP is the pressure drop along the cylinder, μ is the viscosity of the fluid and L is the length of the cylinder.

Increased ΔP

ΔP_i

If the driving force for flow, ΔP, increases, the vessel constricts, increasing the resistance to flow and maintaining the flow rate, Q, relatively constant. Note that increases in ΔP refers to an increase in the magnitude of the pressure drop (i.e., $P_{out}-P_{in}$ becomes more negative).

Decreased ΔP

ΔP_d

If ΔP decreases the vessel dilates, decreasing the resistance to flow and maintaining the flow rate relatively constant.

Fig. 5.2. A demonstration of how the myogenic response flow may provide for a relatively constant flow rate despite variation in blood pressure.

tension remains constant the vessel radius must decrease as pressure rises, as would be qualitatively predicted by the myogenic response. However, a more quantitative analysis of a hypothetical control system based on constant wall tension reveals a potential weakness. Doubling systemic pressure would be predicted to reduce the vessel radius by approximately half to maintain constant wall tension. As flow rate through a vessel increases linearly with pressure drop along the vessel and increases with the radius to the fourth power, the net effect of increasing blood pressure would overcorrect with pressure-induced contractions, decreasing blood flow substantially.

3. Pressure results in deformation of the artery, which may be sensed by smooth muscle cells. Stretch-activated ion channels have often been suggested to explain the coupling of stretch and vascular muscle

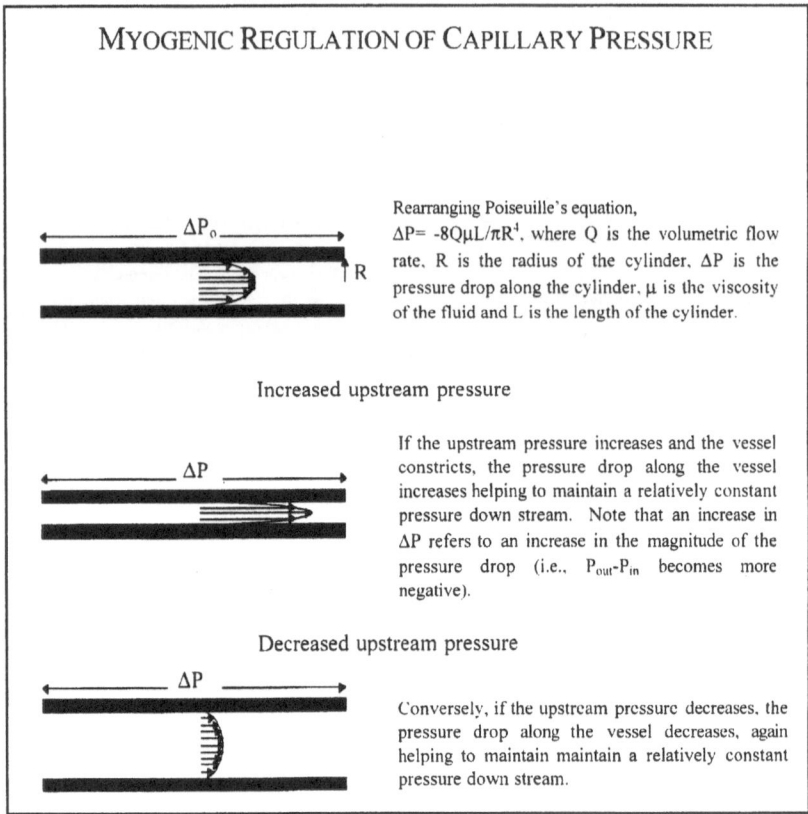

Fig. 5.3. A demonstration of how the myogenic response may provide for a relatively constant capillary pressure despite variation in blood pressure.

contraction. The evidence for stretch-activated ion channels in the myogenic response has been reviewed.[23] Careful attention must be paid to nomenclature here, as stretch-activated ion channels typically (as in the discussion in chapter 6 on mechanotransduction) refer to channels that are *directly* activated by stretch applied to the plasma membrane, and not merely ion channels that are indirectly activated by deformation of the cell or tissue. Although stretch-activated ion channels can be used to explain the transduction of stretch into an ionic event that may mediate the initial contraction, they do not explain the maintenance of contraction following fiber shortening, which would return stretch-activated channels to an unstretched state.

4. An explanation that addresses both the initiation and maintenance of contraction is that higher blood pressure may increase the flow of fluid through the vessel wall, stimulating smooth muscle cells to contract by an unknown mechanism (Fig. 5.4). Though never evaluated experimentally, this hypothesis is attractive in light of mathematical models by Wang and Tarbell predicting that shear stress exerted on the vascular smooth muscle cells by this flow would be roughly the same order of magnitude as that experienced by the endothelium resulting from blood flow. As discussed below, recent studies demonstrate that monolayers of cultured vascular smooth muscle cells respond to these levels of laminar fluid flow by altering their metabolism and proliferation rate. Interestingly, pharmacological inhibitors of G proteins abolish the myogenic response as well as most flow-induced responses in endothelial cells and osteoblasts (see Table 6.2). Additionally, pharmacological inhibitors of protein kinase C abrogate the myogenic response and also have been implicated in several flow-induced responses in endothelial cells.

Chronic effects of mechanical forces on vascular smooth muscle in vivo

Chronic changes in the mechanical environment result in remodeling of the vascular system. The first careful study of this remodeling was done by Thoma more than a century ago.[24] From his observations of the developing vasculosa of the chick embryo, Thoma proposed the following four hypotheses relating vessel development to mechanical forces.

1. Increased blood flow results in increased vessel diameter.
2. Increased blood pressure results in increased vessel wall thickness.
3. Increased axial tension results in increased vessel length.
4. Increased intravascular pressure results in angiogenesis.

Though based on observations of a developing organism, hypotheses one, two and, to a lesser extent, three are consistent with most of the available data for vascular remodeling resulting in altered blood flow in adult organisms. Kamiya and Togawa[25] introduced arterial venule shunts into dogs to increase flow through the vessel up to 10-fold. As predicted by previous work, the vessels increased in diameter to accommodate the increased flow. Interestingly, the increase in diameter was such that the wall shear stress (as calculated by Poiseulle's equation) returned to its original value in all cases except that of highest flow (Fig. 5.5). It is not known if the reason the shear stress did not return to basal levels with the largest increases in blood flow represents a limited ability of the vessel to remodel to the physical stimulus or whether the time from surgical modification to observations (2 months) was too short to permit complete remodeling. In another experimental system, however, up to 30-fold increases in blood flow produced vessel remodeling, resulting in a return to original shear stress after 6 months.[26]

The Effects of Mechanical Forces on Muscle Cells In Vitro

Most in vitro experiments on cultured muscle cells subject the cells to either a single prolonged deformation (static strain) or stretch and relax the cells repetitively (cyclic strain), presumably because these are thought to be the most appropriate models of the forces experienced by cells in vivo. More recently, several studies of the effects of laminar fluid flow on vascular smooth

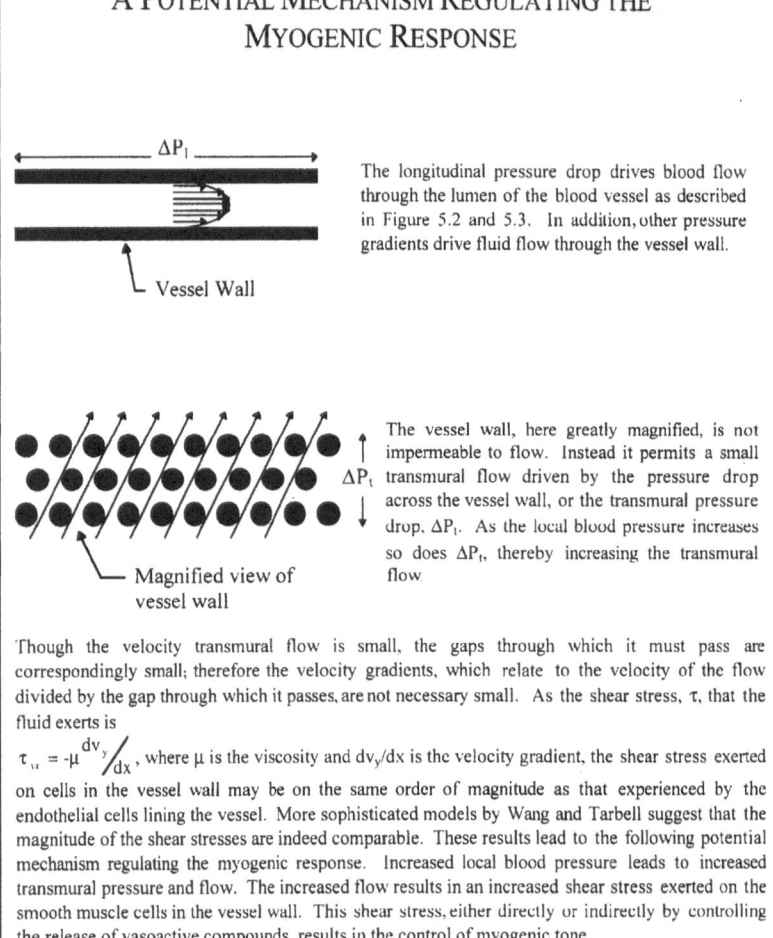

A POTENTIAL MECHANISM REGULATING THE MYOGENIC RESPONSE

ΔP_l

Vessel Wall

The longitudinal pressure drop drives blood flow through the lumen of the blood vessel as described in Figure 5.2 and 5.3. In addition, other pressure gradients drive fluid flow through the vessel wall.

ΔP_t

Magnified view of vessel wall

The vessel wall, here greatly magnified, is not impermeable to flow. Instead it permits a small transmural flow driven by the pressure drop across the vessel wall, or the transmural pressure drop, ΔP_t. As the local blood pressure increases so does ΔP_t, thereby increasing the transmural flow.

Though the velocity transmural flow is small, the gaps through which it must pass are correspondingly small; therefore the velocity gradients, which relate to the velocity of the flow divided by the gap through which it passes, are not necessary small. As the shear stress, τ, that the fluid exerts is

$\tau_{u} = -\mu \, {}^{dv_y}\!/_{dx}$, where μ is the viscosity and dv_y/dx is the velocity gradient, the shear stress exerted on cells in the vessel wall may be on the same order of magnitude as that experienced by the endothelial cells lining the vessel. More sophisticated models by Wang and Tarbell suggest that the magnitude of the shear stresses are indeed comparable. These results lead to the following potential mechanism regulating the myogenic response. Increased local blood pressure leads to increased transmural pressure and flow. The increased flow results in an increased shear stress exerted on the smooth muscle cells in the vessel wall. This shear stress, either directly or indirectly by controlling the release of vasoactive compounds, results in the control of myogenic tone.

Fig. 5.4. An explanation of an hypothetical mechanism controlling the myogenic response.

muscle cells have been performed. The rationale behind these fluid flow experiments is described below.

Skeletal muscle

Isolated skeletal muscle myotubes from embryonic tissue were cultured on silicone membranes for 72 h before the membranes were statically stretched ~10%. The incorporation of radiolabeled α-aminoisobutyric acid and leucine was determined in both stretched and nonstretched cultures. Stretched cultures incorporated approximately 30% more α-aminoisobutyric acid than

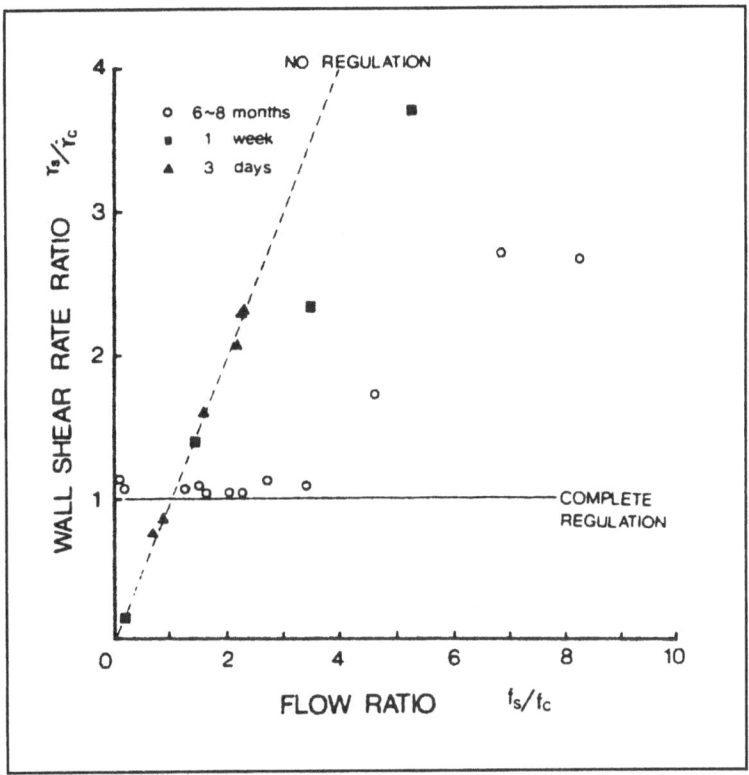

Fig. 5.5. Blood flow through canine carotid arteries was manipulated by the introduction of arterial-venule shunts resulting in regulation of wall shear rate (and shear stress) by adaptive changes in arterial radius. The shear rate ratio (shunt to control) in acute experiments (▲) is on the line of no regulation (i.e., wall shear rate ratio = flow ratio). Chronic data are very near to the line, representing complete regulation (wall shear rate ratio = 1), except those corresponding to very large increases in flow (flow ratio > 4). With permission from Kamiya et al. Am J Physiol 1980; 239:H14-21.

unstretched controls, suggesting that stretch increases amino acid transport in cells. This increase in amino acid incorporation is concurrent with a several-fold increase in incorporation of leucine, an indicator of protein synthesis.[27]

Wright et al examined the effect of cyclic strain on the morphology and proliferation of embryonic skeletal muscle cells cultured on flexible membranes. Cyclic stretch (15-24% elongation at 0.25 Hz) caused the cells to elongate and align perpendicular to the direction of stretch.[28] These authors proposed that the changes in morphology minimized the amount of strain to which the cells were exposed. In the same study, it was observed that ^{3}H-thymidine incorporation, an indicator of DNA synthesis, was greatly increased in stretched cells.

A potential mechanism explaining cyclic stretch-induced proliferation in embryonic muscle cells and hypertrophy by adult skeletal muscle cells in response to loading is suggested by the recent work of Clarke and Feeback regarding human skeletal muscle cells.[29] When exposed to cyclic stretch of 10% and 20%, the amount of basic fibroblast growth factor (bFGF) released into the culture medium increased dramatically (Fig. 5.6, Panel A). In addition, cyclic stretch resulted in an increase in membrane damage as determined by fluorescence by FDxLys,[29] a fixable membrane wound marker (Fig. 5.6, Panel B). Interpreting these results, the authors suggest that cyclic loading results in damage to the plasma membrane, permitting the release of bFGF, which may account for the hypertrophy.

Cardiac muscle

Cardiac myocytes from 1-day-old rats were cultured on a flexible silicone membrane and statically stretched. Stretch upregulated the expression of mRNA for the proto-oncogene *c-fos* in a dose (% stretch) dependent manner. Northern analysis revealed that 5% stretch was adequate to produce detectable upregulation (15% increase), and 20% stretch elicited a maximal response (100% increase). The upregulation of *c-fos* is both rapid and transient, with increases detected after as little as 15 min of static stretch, maximal increases at 30 min, and a return to basal levels after 2 h. These increases in *c-fos* mRNA were followed temporally by an increase in protein content but not thymidine incorporation (Fig. 5.7, Panels A, B and C).[44] Interestingly, the time course of *c-fos* mRNA expression in myocytes exposed to static stretch is virtually identical to that of endothelial cells exposed to steady laminar flow.[30]

Smooth muscle

Of the three types of muscle, smooth muscle has been studied the most extensively in vitro. This interest undoubtedly is motivated by the role of

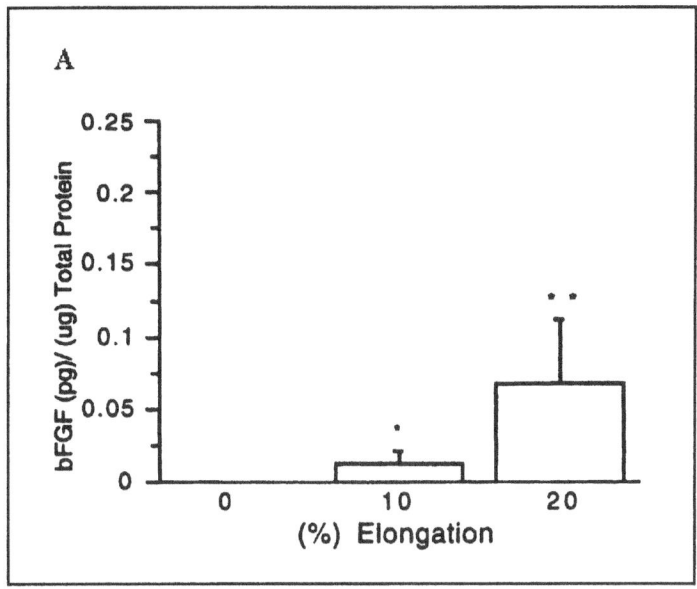

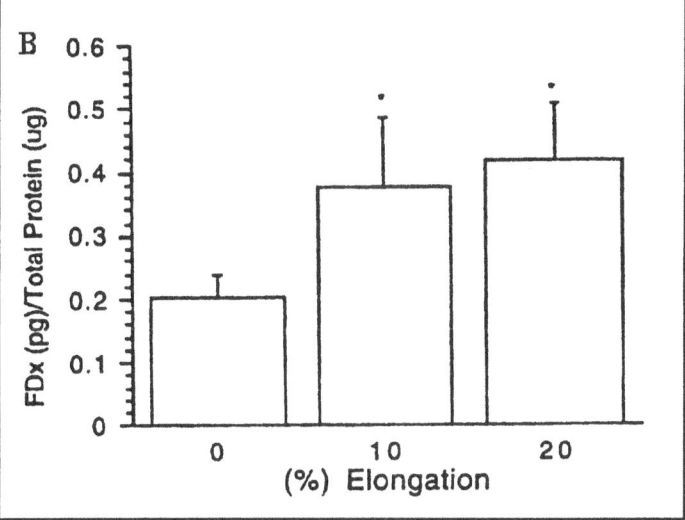

Fig. 5.6. Panel A. The effect of acute mechanical load (10 min) on release of bFGF into the cell culture medium by human skeletal muscle cells. Panel B. Cell lysate-associated FDx, an index of membrane wounding, increases with acute mechanical loading. FASEB J 1996; 10:502-9.

smooth muscle cells in important adaptations to physical forces in vivo (e.g., the myogenic response and blood vessel remodeling). The use of smooth muscle cells from adult animals certainly is facilitated by the ease with which they can be cultured, in contrast to cardiac and skeletal muscle, which are terminally differentiated.

Early ex vivo studies stretched excised sections of aorta[31] to strain the smooth muscles. Using such a system, Leung et al demonstrated that cyclic stretching (10% amplitude at 0.87 Hz) of rabbit aortic smooth muscle cells increased by approximately three-fold the synthesis of total protein; type I, type III, and total collagen; hyaluronate; and chondroitin 6-sulfate.[32] In contrast, in the same experiments dermatan sulfate synthesis increased by 20% while synthesis of chondroitin 4-sulfate was unchanged. Recent reviews of the effects of cyclic stretching of vascular smooth muscle cells in vitro are available.[33-35]

Subsequent in vitro studies utilized smooth muscle cells cultured on a flexible membrane. In one study, smooth muscle cells were cultured on a flexible membrane that was either cyclically stretched (10% amplitude at 0.87 Hz), statically stretched (10% amplitude), or unstretched. Morphometric analysis of the cells revealed increased amounts of rough endoplasmic reticulum in cyclically stretched cells, while static stretch resulted in cytoplasmic degradation and the disappearance of myofilaments.[36]

Later studies were conducted to determine the effect of cyclic stretch on the synthesis of additional proteins. Cyclically stretched and stationary cultures of porcine aortic smooth muscle cells were labeled with ^{35}S-methionine.[33] Proteins from the lysates of these cultures were resolved with two-dimensional gel electrophoresis and quantified with autoradiography and densitometry. Cyclic stretch increased by six-fold the synthesis of desmin and vimentin, two cytoskeletal components, while actin and tropomyosin levels were not affected appreciably. In addition to these identified proteins, the synthesis of more than 9,000 unidentified compounds was quantified. Synthesis of many of these proteins increased, while the synthesis of only three proteins (0.03% of the total number of proteins) decreased.[33] These results contrast with similar studies of endothelial cells in which the synthesis of many proteins either increased or decreased.[33]

In other studies the effect of cyclic stretch (24% maximum amplitude, 10 s stretch, 10 s relaxation for 3 days) on proliferation rate and alignment of cultured porcine aorta cells was determined. Cyclic stretch decreases smooth muscle cell proliferation and ^{3}H-thymidine incorporation. In addition, conditioned medium from stretched smooth muscle cells inhibited the proliferation of quiescent smooth muscle cells.[37]

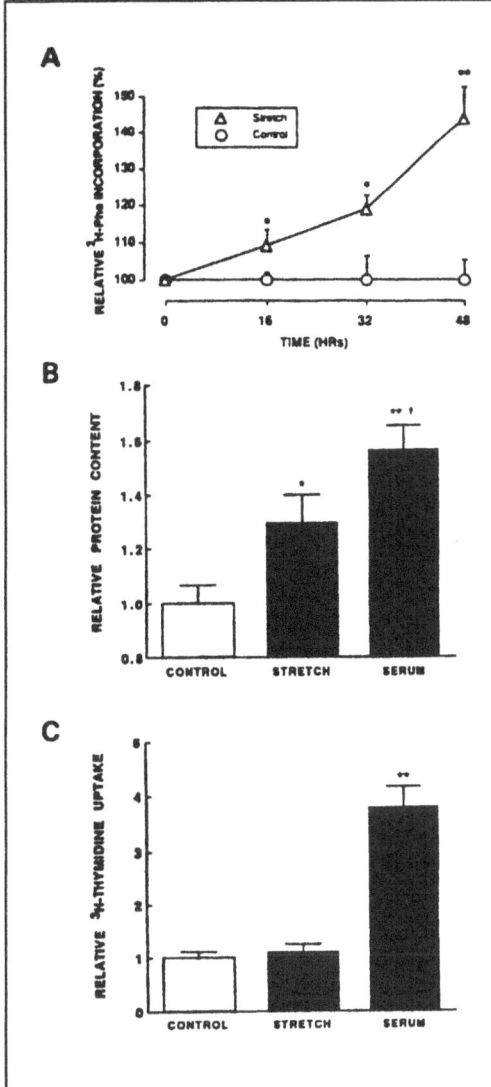

Fig. 5.7. Effect of stretch and serum stimulation on 3H-phylalanine incorporation (A), protein content (B), and 3H-thymidine uptake (C) in cardiac myocytes. A, 3H-phenylalanine incorporation. Myocyte cultures in a serum-free medium were prepared as in Figure 1.B of source article. At each time point, the cpm of 3H was normalized to mean of nonstretched control. Data were also normalized to protein content of the dish (to adjust for the small variability of the cell counts between dishes), although we observed the same results with this normalization. Each data point represents mean + S.E. from four to six samples (*p < 0.05, **p < 0.01 versus control). B, protein content. Cardiac myocytes were stimulated for 48 h. Data were normalized by mean of nonstimulated control. Data were also normalized by DNA content of the dish to normalize the small variability of the cell counts between dishes. Each data represents mean + S.E. from 9 to 10 samples (*p < 0.05, **p < 0.01 versus control +p < 0.05 versus stretch). C, 3H-thymidine uptake. Cardiac myocytes were stimulated for 24 h. For serum stimulation, medium with 20% fetal calf serum was used. Data are expressed as relative cpm/dish normalized to the mean cpm of control cells in each experiment. Data are mean + S.E. from four to five samples (**p , 0.001 versus control). With permission from Sadoshima et al. Molecular characterization of the stretch-induced adaptation of cultured cardiac cells. J Biol Chem 1992; 267)15):10551-10560.

In addition to the above studies, which investigated responses to cyclic strain occurring over a period of a day or more, changes initiated within seconds following stretch have been investigated. Davis and coworkers fluorometrically measured intracellular calcium concentrations in cultured vascular smooth muscle cells immediately following a step increase in length. They observed a rapid increase in intracellular calcium after several seconds. If the cell remained stretched, intracellular calcium remained elevated for at least several minutes. The magnitude of the increase in calcium was sigmoidally related to the percentage change in cell length. Increasing the extracellular concentration of calcium to 10 mM potentiated the response, while chelating extracellular calcium with EGTA abolished the response, suggesting that the increase in intracellular calcium was primarily due to calcium influx rather than the release of intracellular stores. Based on pharmacological studies, the authors proposed a credible series of events that may link smooth muscle cell stretch and the initial contraction observed in the myogenic response in vivo, though, as mentioned above, it is difficult to explain the maintenance of myogenic tone by an acute stimulus (e.g., stretch).

In addition to stretch, smooth muscle cells respond to fluid flow. In one study, bovine arterial smooth muscle cells were seeded on fibronectin-coated polystyrene and allowed to adhere for 48 h before being exposed to flow or stationary conditions.[38] Flow for 24 h inhibited cell proliferation and ^{3}H-thymidine incorporation, a result which has been corroborated in human aortic smooth muscle cells (Fig. 5.8).[39] The extent of inhibition depended on the applied shear stress within the range studied (3, 6 and 9 dyn/cm²). The inhibition of proliferation persisted for at least 24 hours after the cessation of flow. Flow cytometry analysis of cell-cycle distribution demonstrated a lower number of cells in S phase in cultures exposed to flow compared to stationary controls.[38] Flow also caused a two-fold stimulation of cell-associated protein synthesis.[40] Interestingly, in light of the observed inhibition of proliferation, smooth muscle cells exposed to fluid flow released higher quantities of mitogens, including a platelet-derived growth factor (PDGF)-like substance, as detected by immunological testing.[40] Conditioned medium from smooth muscle cells subjected to flow produced a three-fold greater increase in ^{3}H-thymidine incorporation in Swiss 3T3 cells than medium conditioned by smooth muscle cells under stationary conditions. Addition of an anti-PDGF antibody to medium conditioned by smooth muscle cells exposed to fluid flow inhibited the increase in ^{3}H-thymidine incorporation in Swiss 3T3 cells by 30%, while an anti-bFGF antibody decreased proliferation by 60%.[41]

Laminar fluid flow also modulates the production of prostaglandins by cultured vascular smooth muscle cells (Fig. 5.9, Panels A and B). Application of laminar flow with a maximum wall shear stress of 5 or 20 dyn/cm² increases

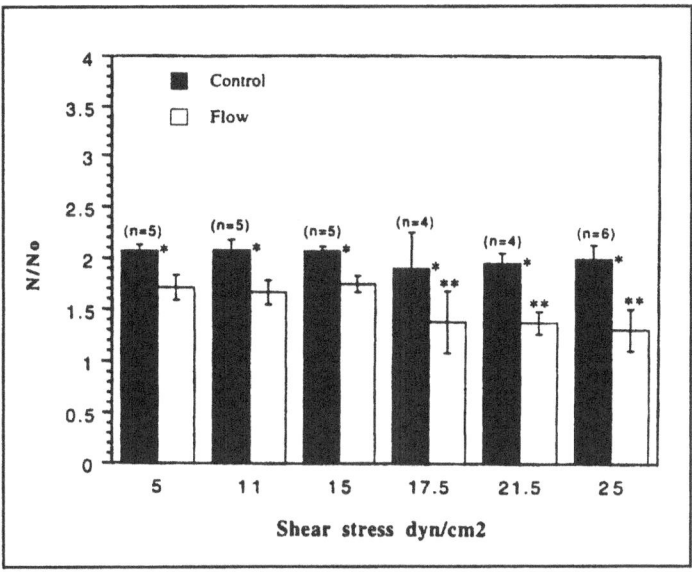

Fig. 5.8. Effects of shear stress on the growth rate of cultured human aortic smooth muscle cells (hASMC). N indicates the number of cells per sq. centimeter at the end of a 24-h experiment, while No indicates the number of cells per sq. centimeter at the beginning of an experiment. Asterisks indicate significant difference from the respective stationary culture cell number; double asterisks indicate significant difference from the cell number at 5, 11 and 15 dyn/cm² shear stress. With permission from Papadaki M et al. Effects of shear stress on the growth kinetics of human aortic smooth muscle cells in vitro. Biotech Bioeng 1996; 50:555-61. Copyright © 1996. Reprinted by permission of John Wiley & Sons, Inc.

the production of both PGE_2 and prostacyclin as measured by the stable break-down product 6-keto-PGF_1, though notable differences in the responses are evident. Enhanced PGE_2 release occurs after more than 2 hrs while increases in prostacyclin are evident within 1 hr. The magnitude of the PGE_2 response is dependent on the magnitude of the shear stress applied, while 0.5, 1, and 20 dyn/cm² all elicit similar cumulative prostacyclin release after 6 hr.[42]

While the metabolic response of cultured vascular smooth muscle cells resembles that of cultured endothelial cells in many ways (inhibition of proliferation, synthesis of mitogens), their responses to flow are not identical. Rabbit aortic endothelial cells increased their release of ATP 14-fold as a result of increasing the perfusion rate from 0.5 ml/min to 3.0 ml/min. However, smooth muscle cells exposed to identical flow conditions did not increase their release of ATP above basal levels.[43]

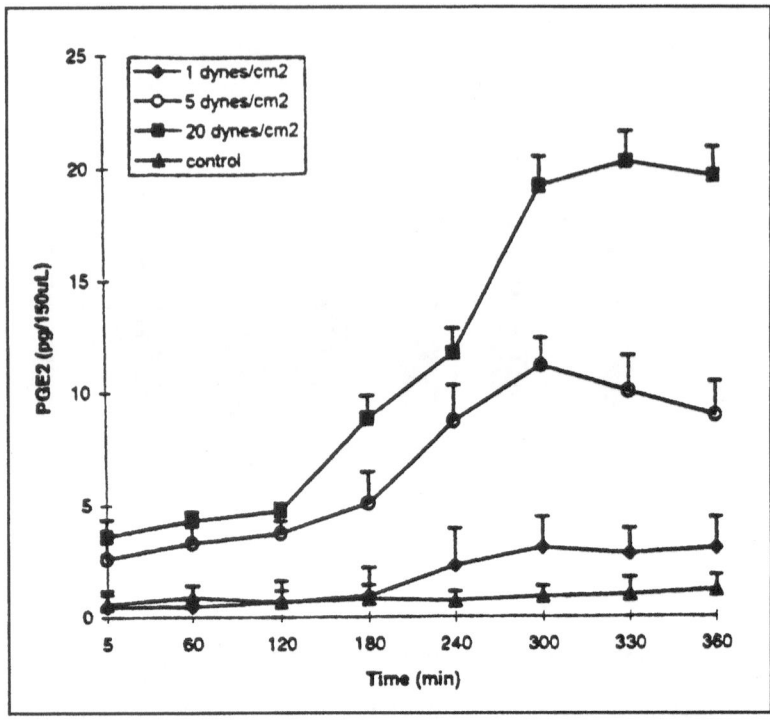

Fig. 5.9. A (this page), PGE_2 cumulative concentration versus time for three levels of shear stress: 1 dyn/cm² (n = 6); 5 dyn/cm² (n = 6); and 20 dyn/cm² (n = 6). Controls (no shear) (n = 18). The 5 and 20 dyn/cm² levels of shear produced significantly different quantities of PGE_2 compared to controls and to each other (p < 0.05) at 6 hours. Data is presented as mean ± SEM. B (opposite page), 6 keto-$PGF_1\alpha$, the stable breakdown product of prostacyclin, cumulative concentrations versus time for three levels of shear stress: 0.5 dyn/cm² (n = 6); 1 dyn/cm² (n = 6); and 20 dyn/cm² (n = 6). Controls (no shear) (n = 18). No significant difference in concentrations among the three levels of shear (p > 0.1) at 6 hours, but all were significantly different from controls (p < 0.05). Data is presented as mean ± SEM. With permission from Alshihabi SN. Biochem Biophys Res Commun 1996; 224:808-14.

Summary

Skeletal, cardiac, and smooth muscle cells are exposed to, and respond to, an array of mechanical forces in vivo. Most in vitro experiments on cultured muscle cells either subject the cells to a single prolonged deformation (static strain) or stretch and relax the cells repetitively (cyclic strain) in an attempt to simulate in vivo conditions. Strain induces elevated gene

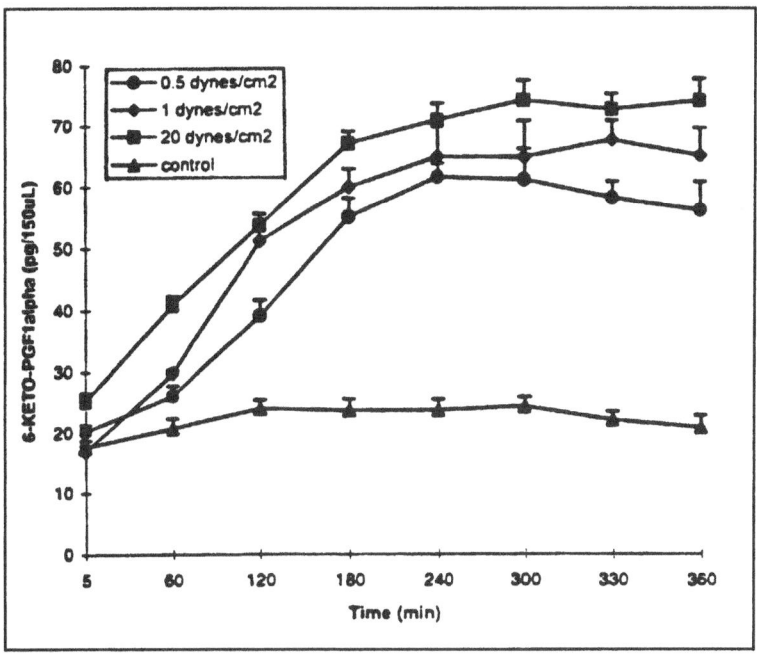

Fig. 5.9B. (see legend, opposite page).

expression, increases protein synthesis, and alters cellular morphology. Recently, several studies of the effects of laminar fluid flow on vascular smooth muscle cells have suggested that fluid flow inhibits cellular proliferation and affects the synthesis of prostaglandins.

References

1. Eisenberg BR. Quantitative ultrastructure of mammalian skeletal muscle. In: Handbook of Physiology. Skeletal Muscle 1983; Am Physiol Soc: Bethesda, MD. p. 73-112.
2. Thomason DB et al. Time course of soleus muscle myosin expression during hindlimb suspension and recovery. J Appl Physiol 1987; 63:130-7.
3. Booth FW, Kirby CR. Changes in skeletal muscle gene expression consequent to altered weight bearing. Am J Physiol 1992; 262:R329-332.
4. Thomason DB, Biggs RB, Booth FW. Altered protein metabolism and unchanged B-myosin heavy chain mRNA in unweighted soleus muscle. Am J Physiol 1989; 257:R300-5.
5. Goldspink DF. Th influence of immobilization and stretch on protein turnover of rat skeletal muscle. J Physiol Lond 1977; 282(264):267-82.

6. Wong TS, Booth FW. Protein metabolism in rat tibialis anterior-muscle after stimulated chronic eccentric exercise. J Applied Physiol 1990; 69:1718-24.

7. Pette D, Dusterhoft S. Altered gene expression in fast-twitch muscle induced by chronic low-frequency stimulation. Am J Physiol 1992; 262:R333-8.

8. Salmons S, Vrbova G. The influence of activity on some contractile characteristics on mammalian fast and slow muscles. J Physiol Lond 1969; 201:535-549.

9. Puri PL et al. The molecular basis of myocardial hypertophy. Ann Ital Med Int 1994; 9(3):169-5.

10. Yamazaki T, Komuro I, Yazaki Y. Molecular mechanism of cardiac cellular hypertrophy by mechanical stress. J Mol Cell Cardiol 1995; 27(1):133-40.

11. Gorza L et al. Isomyosin distribution in normal pressure overloaded rat ventricular myocardium. An immunohistochemical study. Circ Res 1981; 49:1003-9.

12. Lompre AM et al. Myosin isoenzyme redistributes in chronic heart overload. Nature 1979; 282:105-7.

13. Schwartz K, Boheler KR, Bastie D. Switches in cardiac muscle gene expression as a result of pressure and volume overload. Am J Physiol 1992; 262:R364-6.

14. Bayliss WM. On the local reactions of the arterial wall to changes of internal pressure. J Physiol Lond 1902; 28:220-31.

15. Harder DR. Pressure-induced myogenic activation of cat cerebral arteries is dependent on intact endothelium. Circ Res 1987; 60:102-7.

16. Katusic ZS, Shepherd JT, Vanhoutte PM. Endothelial-dependent contraction to stretch in canine basilar arteries. Am J Physiol 1987; 252:H671-3.

17. Bulow A. Myogenic activity in isolated resistance arteries from skeletal muscle of the rat. Blood Vessels 1991; 28:278-9.

18. Falcone JC, Davis MJ, Meininger GA. Endothelial independence of the myogenic response in skeletal muscle arterioles. Am J Physiol 1991; 260:H130-5.

19. Hwa JJ, Bevan JA. Stretch-dependent (myogenic) tone in rabbit ear resistance arteries. Am J Physiol 1986; 250:H87-95.

20. Kuo L, Chilian WM, Davis MJ. Coronary arteriolar myogenic response is independent of endothelium. Circ Res 1990; 66:860-6.

21. MacPherson RS, McLeod LJ, Rasiah RL. Myogenic response of isolated pressurized rabbit ear artery is independent of endothelium. Am J Physiol 1991; H779-84.

22. McCarron JG, Osol G, Halpern W. Myogenic responses are independent in the rat pressurized posterior cerebral arteries. Blood Vessels 1989; 26:415-9.

23. Meininger GA, Davis MJ. Cellular mechanisms involved in the vascular myogenic response. Am J Physiol 1992; 263:H647-59.

24. Thoma R. Untersuchagen uberdie Histogenese und Histomechanik des Gefassystems. Stuttgart: Enke, 1893.

25. Kamiya A, Togawa T. Adapted regulation of wall shear stress to flow changes in the canine carotid artery. Am J Physiol 1980; 239:H14-21.

26. Zarins CK et al. Shear stress regulation of artery lumen diameter in experimental atherogenesis. J Vasc Surg 1987; 5:413-20.

27. Vandenburgh H, Kaufman S. In vitro model for stretch-induced hypertrophy pf skeletal muscle. Science 1979; 203:265-8.

28. Wright E, MacMurray R, Banes A. Alignment rates of skeletal myocytes subjected to cyclic stretch in vitro. J Cell Bio 1988; 107:453a.

29. Clarke MS, Feeback DL. Mechanical load induces sarcoplasmic wounding and FGF release in differentiated human skeletal muscle cultures. FASEB J 1996; 10:502-9.

30. Hsieh HJ, Li NQ, Frangos JA. Pulsatile and steady flow induces c-fos expression in human endothelial cells. J Cell Physiol 1993; 154:143-51.

31. Leung DYM, Glagov S, Matthews MB. A new in vitro system for studying cell response to mechanical stimulation. Exp Cell Res 1977; 109:285-98.

32. Leung DYM, Glagov S, Matthews MB. Cyclic stretching stimulates synthesis of matrix components by arterial smooth muscle cells in vito. Science 1976; 191:475-7.

33. Mills I, Cohen CR, Sumpio BE. Cyclic strain and vascular cell biology. In: Sumpio BE, ed. Hemodynamic Forces and Vascular Cell Biology. Austin: R.G. Landes Company, 1993; 66-89.

34. Isales C, Rosales O, Sumpio BE. Mediators and mechanism of cyclic strain and shear stress-induced vascular responses. In: Sumpio BE, ed. Hemodynamic Forces and Vascular Cell Biology. Austin: R.G. Landes Company, 1993.

35. Patrick CW, McIntire LV. Shear stress and cyclic strain modulation of gene expression in vascular endothelial cells. Blood Purif 1995; 13(3-4):112-24.

36. Sottiurai V et al. Morphological alteration of cultured arterial smooth muscle cells by cyclic stretching. J Surg Res 1983. 35:490-7.

37. Sumpio BE, Banes AJ. Response of porcine aortic smooth muscle cells to cyclic tensional deformation in culture. J Surg Res 1988; 44:696-701.

38. Sterpetti AV et al. Modulation of arterial smooth muscle cell growth by haemodynamic forces. Eur J Vas Surg 1992; 6(1):16-20.

39. Papadaki M, McIntire LV, Eskin SG. Effects of shear stress on growth kinetics of human aortic smooth muscle cells in vitro. Biotech Bioeng 1996; 50:555-61.

40. Sterpetti AV et al. Shear stress modulates the proliferation rate, protein synthesis, and mitogenic activity of arterial smooth muscle cells. Surgery 1993; 113(6):691-9.

41. Sterpetti AV et al. Shear stress influences the release of platelet derived growth factor and basic fibroblast growth factor by arterial smooth muscle cells. Eur J Vasc Surg 1994; 8(2):138-42.

42. Alshihabi SN, Chang YS, Frangos JA et al. Shear stress-induced relsease of PGE_2 and PGI_2 by vascular smooth muscle cells. Biochem Biophys Res Commun 1996; 224:3, 808-14.

43. Bodin P, Bailey D, Burnstock G. Increased flow-induced ATP release from isolated vascular endothelial cells but not smooth muscle cells. Br J Pharmacol 1991; 103(1):1203-5.

44. Sadoshima JI, Takahashi T, Jahn L, Isumo S. Role of mechano-sensitive ion channels, cytoskeleton, and contractile activity in stretch-induced immediate-early gene expression and hypertrophy of cardiac myocytes. Proc Natl Acad Sci USA 1992; 89:9905-9.

CHAPTER 6

Mechanotransduction

One of the most intensely debated questions arising from the study of cellular responses to physical forces is how cells sense these mechanical stimuli and convert them into electrical, chemical or biochemical responses, the process known as mechanotransduction. Systems in which mechanotransduction is essential include the senses of hearing and touch as well as baroreception, the ability of the body to perceive blood pressure. In all three cases, mechanical forces move or deform specialized sensory structures such as hair cells (hearing); Pacinian corpuscles, Meissner's corpuscles, and free nerve endings (touch); or baroreceptors. These physical stimuli lead to altered membrane potential, which in turn triggers the initiation of action potentials. In these examples it is widely accepted that the mechanosensitive elements are ion channels whose probability of being open is affected by physical deformation (reviewed by French[1] and Sachs[2,3]). Once a physical stimulus is converted to an electrochemical signal, the transfer and amplification of the signal follows classic signal transduction pathways often utilized by agonists.

As has already been emphasized in this text, the ability of cells to sense and respond to physical forces is not limited to specialized sensory cell types but is exhibited by most cell types studied. Most types of animal cells studied thus far respond to physical forces, even those not thought to be subjected to large physical forces in vivo, suggesting that components common to most cells are adequate for mechanotransduction, although the presence of specialized mechanotransducing elements in some cell types cannot be excluded.

Since physical forces may be accompanied by modifications of the electrochemical environment, it is important to determine whether cells respond directly to forces or to the electrochemical changes they induce. The example of fluid flow over the surface of a cell illustrates how a physical force, shear stress, may be accompanied by changes in the chemical and electrical environment. Although flow across the surface of a cell occurs in numerous instances in vivo and in cell culture, laminar flow over a confluent monolayer of endothelial cells in a parallel-plate flow chamber will be discussed for two reasons. First, the well-defined geometry of this system allows mathematical

Mechanical Forces: Their Effects on Cells and Tissues, by Keith J. Gooch and Christopher J. Tennant. © 1997 Landes Bioscience.

modeling of the physical forces and the attendant changes in the electrochemical environment; second, experimental data utilizing this system are plentiful.

Variations in the chemical environment

Fluid flow increases the transport of agonists and other compounds to and from the surface of cells. If these agonists are degraded at the cell surface, as ATP is, flow will result in an increase in their concentration at the cell surface, suggesting that flow-induced responses may be mediated by increases in agonist concentrations rather than by shear stress or other mechanical forces. Consistent with this hypothesis, many of the acute responses of endothelial cells to flow and ATP, such as increased inositol hydrolysis, intracellular Ca^{2+} concentration and nitric oxide (NO) production, are similar. In addition, some investigators report that flow-induced increases in intracellular Ca^{2+} concentrations are observed when medium containing ATP is used, but not when ATP-free medium is used.[4,5]

Although the above data are consistent with the hypothesis that flow effects are mediated by local increases in agonists such as ATP, a substantial body of data disputes this contention. Some researchers have reported the elevation of intracellular Ca^{2+} in endothelial cells exposed to flow of ATP-free medium. In addition, there are many reports of other flow-induced phenomena (modified ET-1 production,[6] elevated NO production[7,8] and cGMP concentration[7,8]) in the presence of medium lacking both ATP and serum.

It is possible to argue that an agonist other than ATP degrades at the surface of the cell and mediates the observed flow-induced responses. One candidate is bradykinin, which stimulates endothelial cells to increase intracellular Ca^{2+} and NO production by binding to B2 receptors on the plasma membrane. Bradykinin is degraded by angiotensin converting enzyme on the surface of the cell. The B2 receptor antagonist Hoe 140, however, does not inhibit the flow-induced production of NO by endothelial cells, suggesting that flow-enhanced transport of bradykinin to the cell surface does not mediate the response.[9] Potentially, other discovered and/or undiscovered agonists could mediate the response. By using well-defined media such as serum-free medium with all potential agonists such as ATP excluded, it is possible to minimize the source of external agonists. However, since cells may release autocrine and paracrine factors, the local concentration of which would be altered by fluid flow, it is difficult to completely reject the hypothesis that flow stimulates cells by modulating the surface concentrations of those factors based on the above studies.

Mathematical models provide researchers with useful insights into how to decouple the physical and chemical effects of fluid flow. These models predict that the concentration of a diffusible compound at the surface of the cell is dependent on the shear rate but not the physical force of shear stress.[10] Figure 6.1 describes mathematically the difference between shear rate and shear stress. Conceptually, shear stress is a physical force exerted by the moving fluid while shear rate is a description of the relative motion of the fluid. Because the surface concentration of a compound depends on shear rate rather than shear stress, it is possible to evaluate the hypothesis that flow changes the concentration of any compound at the surface of the cell, resulting in the observed flow-induced changes.

By supplementing the perfusion medium with dextran, thus increasing its viscosity while maintaining constant flow rates, Melkumyants et al[11] demonstrated that vasodilation of excised arteries correlates with shear stress rather than shear rate. Complementary studies of cultured cells also show that flow-induced production of NO by endothelial cells[12,13] and cAMP production by osteoblasts[14] are dependent on shear stress rather than shear rate (Fig. 6.2). Taken together, these data suggest that fluid flow may mediate certain biological responses such as increases in intracellular Ca^{2+} by increasing the concentration of ATP at the surface of the cell. Many other acute and chronic responses appear to be mediated by other mechanisms and caused directly by the shear stress applied to the cell.

Forces other than shear stress are associated with changes in the electrochemical environment. Increases in pressure, for example, may be associated with increased concentrations of dissolved gases. Increases in dissolved oxygen or carbon dioxide are likely to modify cellular function independent of the applied pressure. Physical forces can have direct biological effects on cells. However, when considering data it is important to determine whether proper control experiments have been conducted, so effects due to mechanical forces can be distinguished from those due to changes in the chemical environment.

Variations in the electrical environment

The cell's glycocalyx has a net negative charge that attracts and tightly binds a layer of cations. A layer of anions in the adjacent fluid is loosely associated with these cations. Fluid motion carries along the mobile anions, but does not move the tightly bound cations; this current results in a net movement of charges often referred to as a streaming potential.[15] Researchers have hypothesized that flow-induced responses are mediated by streaming potentials. Streaming potential, however, like flow-enhanced transport of agonists, is dependent on the shear rate. The experiments with excised arteries[11] and osteoblasts[14] in which the viscosity of the perfusion medium was modified would argue against this mechanism. In addition, Hung et al[16] have shown that stationary cultures of osteoblasts subjected to a potential calculated

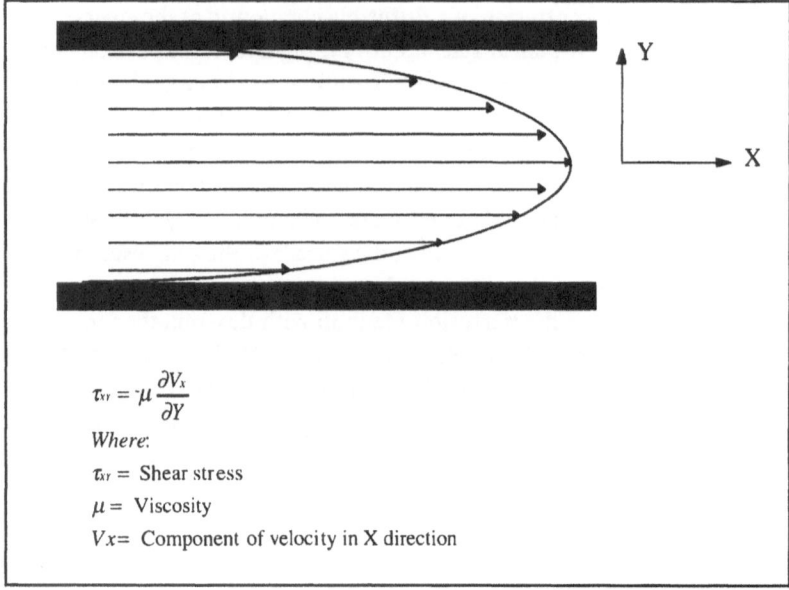

Fig. 6.1. Laminar fluid flow between two parallel plates generates a parabolic velocity profile. The shear stress, τ, generated by the fluid motion is related to the velocity gradient (shear rate) by the viscosity of the fluid.

to simulate the streaming potential experienced by cultures subjected to flow did not increase intracellular calcium concentrations in osteoblasts. Moreover, applying a potential calculated to negate the streaming potential did not inhibit the flow-induced calcium increases in osteoblasts, arguing that at least some of the flow-induced responses in osteoblasts are not dependent on streaming potentials.

Potential mechanisms of mechanotransduction

It appears that many responses elicited by physical forces result from cells directly sensing those forces rather than the subsequent changes in electrochemical environment. An obvious question is how cells sense physical forces. Though many hypotheses have been advanced, no single hypothesis enjoys the support of a substantial majority of researchers in this field, perhaps because no one hypothesis is consistent with the majority of experimental data. This may be because different cell types utilize different mechanisms for mechanotransduction. On the other hand, this may reflect deficiencies in the proposed hypotheses. What follows is a presentation of the major theories of mechanotransduction. Each will be presented as a hypothesis and will be

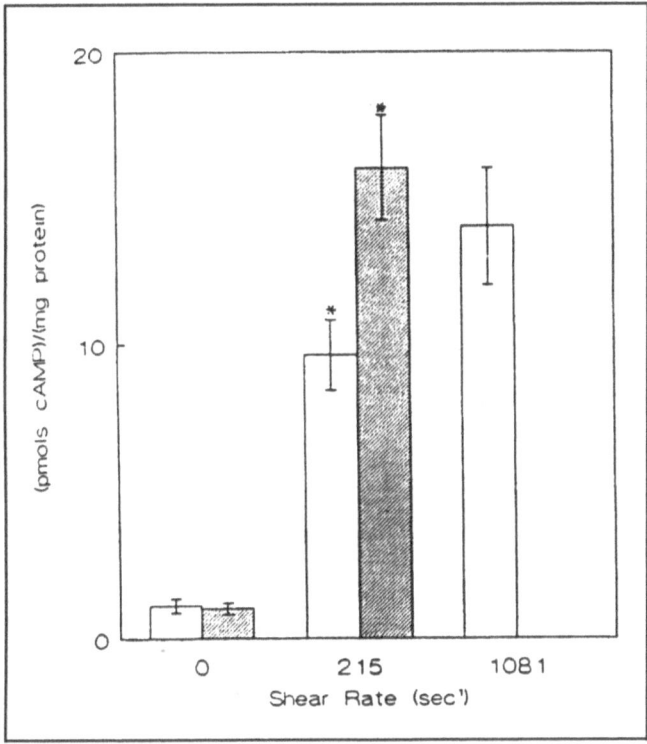

Fig. 6.2. Levels of cAMP were measured from cultured osteoblasts which were subjected to either stationary or flow conditions with either normal (1 centipoise, open bars) or high-viscosity (5 centipoise, shaded bars) medium. The high-viscosity medium resulted in a five-fold greater shear stress than normal medium at the same shear rate. Note that the effect of the high viscosity medium was the same as a five-fold increase in shear rate, suggesting that the osteoblasts are responding to shear stress and not shear rate. From Reich KM, Gay CV, Frangos JA. Fluid shear stress as a mediator of osteoblast cyclic adenosine monophosphate production. Reich KM et al. Fluid shear stress as a mediator of osteoblast cyclic adenosine monophosphate production. J Cell Physiol 1990; 143:100-4. Copyright © 1990. Reprinted by permission of Wiley-Liss, Inc., a subsidiary of John Wiley & Sons, Inc.

followed by data consistent with the hypothesis. Next, major criticisms of the hypothesis and data inconsistent with it will be presented, followed by descriptions of experiments that could be conducted to critically evaluate the hypothesis. An overview of this information is presented in Table 6.1.

Most attempts to locate potential force transducers have focused on two regions of the cell: the plasma membrane and the cytoskeleton. Most forces

Table 6.1. An overview of the three major hypotheses of mechanotransduction.

Proposed Mechano-transducers	Subcellular Location of Mechanotransducer	Major Support and Consistent Data	Major Criticisms and Inconsistent Data	Proposed Experiments and General Comments
Mechanically Activated Ion Channels (MAC)	Plasma Membrane	• Sensitive measurements of channel activation resulting from suction applied to plasma membrane. • Provides a simple relationship between an applied force and a biochemical response, especially in systems such as hearing, touch, and baroreception where the biochemical response is electrical in nature. • Gd^{3+}, an inhibitor of some classes of MAC, abrogated stretch-induced increases in intracellular Ca^{2+} in endothelial cells.[29]	• Hyposmotic swelling, stretch, prodding, and elevation of intracellular pressure failed to elicit the neuronal macroscopic currents predicted by single-channel studies in *Lymnaea stagnalis*.[34] • Though cardiac myocytes exhibit MAC in patch-clamp studies whose activity is abolished by Gd^{3+}, this ion does not inhibit stretch-induced upregulation of *c-jun*, Erg-1, *c-fos*, or *c-myc* expression.[35]	• Though MAC, a well documented phenomena in patch-clamp studies from many cell types, provides a conceptually attractive mechanism connecting applied mechanical forces and biochemical responses, additional experiments documenting the involvement of these channels in mediating mechanically activated responses are required. • The development of specific potent inhibitors of these channels would greatly aid understanding of their significance.

G Protein Linked Receptors or G Proteins	Plasma Membrane	• Many flow-induced responses inhibited by G protein inhibitors and stimulated by G-protein activators (Table 6.2). • Fluid flow (shear stress) stimulates G-protein activation in endothelial cells and liposomes, suggesting direct activation of G proteins by physical forces.	• Some flow-induced responses are not abrogated by G-protein inhibitors, suggesting that other mechanisms of mechano-transduction may be involved.	• Pharmacological studies clearly demonstrate that G proteins are involved in the signal transduction pathways mediating many mechanically-induced responses. It is unclear, however, whether the G proteins are the actual mechanotransducers or if they are involved downstream of the mechanotransducers.
Cytoskeleton	Plasma Membrane and Intracellular	• Mechanical forces are transmitted to and supported by the cytoskeleton, making it a likely mechanotransducer.[46] • Actin depolymerization and elevated ET-1 release are temporarily related in endothelial cells exposed to flow; pharmacological studies suggest a causative relationship.[47] • Experiments indicate that tension on mitotic spindles regulate progression to anaphase during cell division.	• Production of NO by endothelial response to shear stress is not dependent on intact actin or microtubule filaments. • Stretch-induced gene expression in cardiac myocytes is not dependent on intact microtubules or microfilaments.	• Pharmacological studies provide evidence that the cytoskeleton mediates some, but not all, mechanically induced responses. • Several studies reporting a role of the cytoskeleton in mechanotransduction focus on the biomechanical properties of the cell and fail to investigate the role of the cytoskeleton in regulating biological responses to physical forces.

first act directly on the plasma membrane, making it a logical place to look for potential mechanotransducers. Alternatively, since much of the force applied to the plasma membrane is transferred to the cytoskeleton, it too is suspected of housing potential mechanotransducers (Fig. 6.3).

Mechanically Activated Ion Channels

Presentation and consistent data

Ion channels that increase their probability of opening in response to suction applied to the plasma membrane were first identified in embryonic chick skeletal muscle by Guharay and Sachs.[17] The stretch-activated ion channels initially isolated were nonselective cation channels. Additional stretch-activated ion channels were isolated that were selective to K^+, Cl^- and Ca^{2+} as well as those that were nonselective or poorly selective between cations. Another type of channels, the stretch-inactivated ion channels, decrease their permeability in response to mechanical forces. Figure 6.4 illustrates graphically the opening probability of stretch-activated and stretch-inactivated K^+ channels observed in a snail neuron as a function of applied suction. These channels form a broad phenomenological class of mechanically activation channels (MAC). MAC have been identified in more than 30 systems, including prokaryotes, fungi, plants, protozoa, invertebrates and vertebrates including fish, amphibians, birds and mammals (reviewed by Morris[18]).

The hair cells responsible for hearing are possibly the most sensitive mechanotransducers in the human body. The proposed mechanisms by which these specialized columnar epithelial cells sense physical forces has been reviewed by Hudspeth[19] and Pickles and Corey.[20] The current theory regarding mechanotransduction by hair cells is as follows. The mechanosensitivity of each stereocilium is enhanced by its connection by a tip link to its nearest neighbors,[21] forming chains of interconnected stereocilia of progressively increasing length. The deflection of a stereocilium toward its taller neighbor results in the activation[22] of the one or two ion channels[19] located on the tip of the stereocilia that are selective for alkali cations and, to a lesser extent, Ca^{2+}.[23,24] The putative MAC has not been identified or isolated, however, and specific pharmacological inhibitors for the channel are not known.

MAC also have been proposed to mediate baroreception. It is suggested that stretch leads to activation of cation channels, which triggers depolarization of the cell and subsequent generation of action potentials.[25] The trivalent lanthanide gadolinium (Gd^{3+}) has been reported to be an inhibitor of certain stretch-activated channels[26] but did not affect the discharge response of rat baroreceptors at varying pressures.[27] Since it is not known whether Gd^{3+}

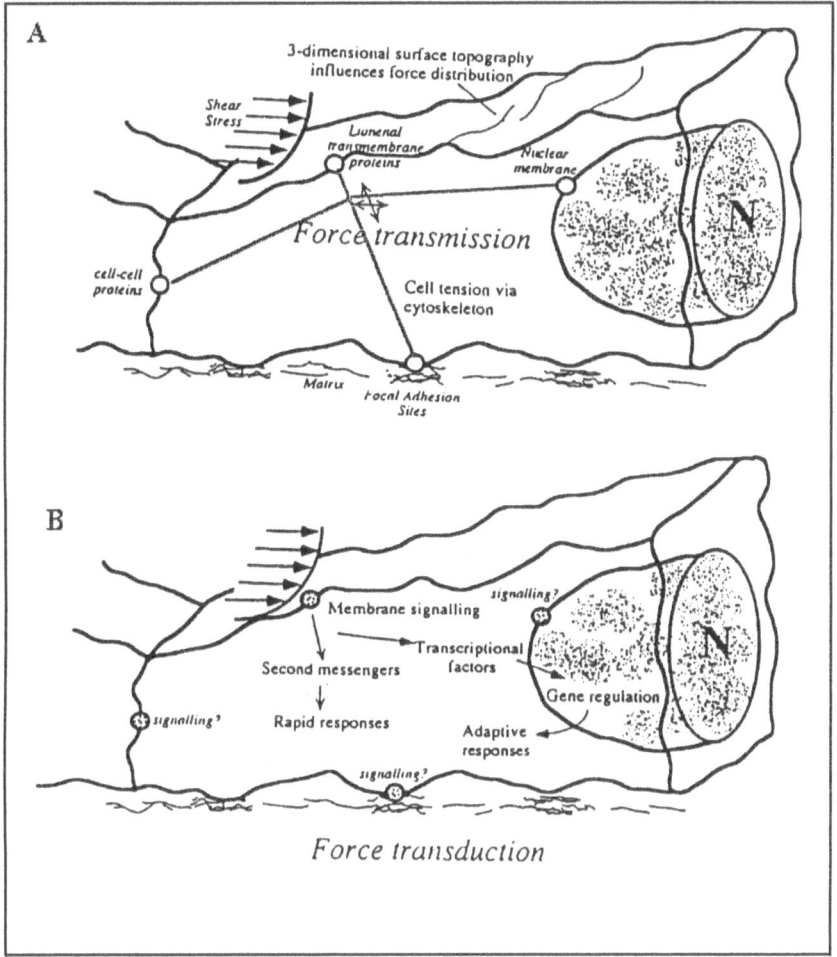

Fig. 6.3. The major hypotheses describing mechanotransduction focus on cytoskeletal (Panel A) or plasma membrane (Panel B) bound elements. From Davies PF. Flow-mediated endothelial mechanotransduction. Physiol Rev 1995; 75(3):519-60, with permission.

inhibits the stretch-activated channels in this system, it is difficult to fully interpret these data.

Additional evidence implicating stretch-activated ion channels in a potentially physiological response is the work of Naruse et al with cultured endothelial cells.[28] Following a 3 s stretch, which increased the surface area of the cell by up to 80%, intracellular Ca^{2+} concentration increased by as much as 175 nM. These increases were abolished by incubating the cells in Ca^{2+}- free medium prior to and during the stretch. Pretreatment of the cells with 10 μM Gd^{3+} almost completely abolished the stretch-induced increase in intracellular

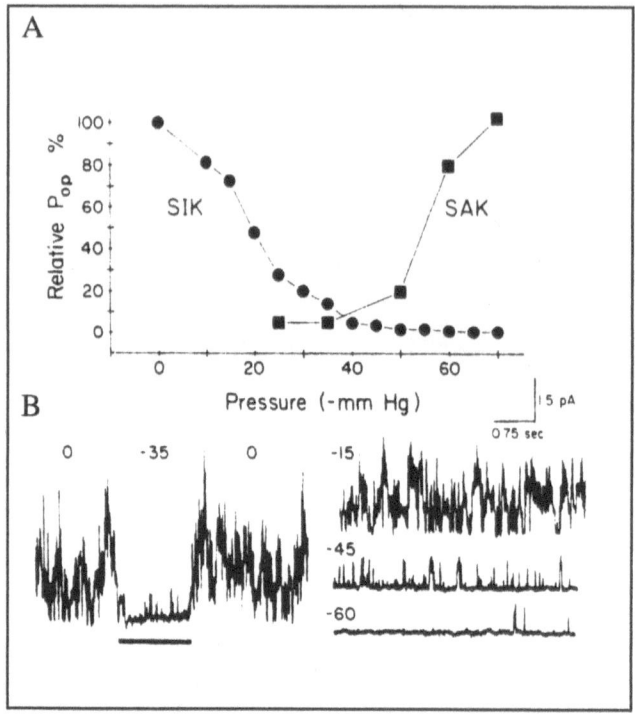

Fig. 6.4. The opening probability of stretch-activated and stretch-inactivated ion channels in snail neuron (Panel A). Single channel conductances for snail neuron at varying applied pressures (Panel B). From Morris C. Mechanosensitive ion channels. J Membrane Biol; 113:93-107. © Springer-Verlag New York Inc., with permission.

Ca^{2+}, suggesting the involvement of a Gd^{3+}-sensitive pathway, perhaps stretch-activated ion channels.

It is very appealing to speculate that mechanotransduction, at least in electrically excitable cells, is mediated by MAC, since these channels would provide a direct coupling between stimulus and response. Observations that mechanical deformation affects the rate and rhythm of the heart led to subsequent studies revealing that stretch initiates and alters action potentials, making the heart a prime candidate for a physiological system in which mechanically activated ion channels may act as mechanotransducers.[29] Sachs incorporated parameters for the reverse potential gating (probability of channel opening as a function of stretch), channel density and conductance of stretch-activated ion channels available from the literature into a computer simulation of electrophysiological performance of stretched heart cells.[30] The output from these simulations was compared to experimental results and the parameters were modified to improve agreement between the simulation predictions and experimental data. The optimized values for most of these parameters fell within the range of those measured experimentally, though the reverse potential cannot be reconciled with published data. This work is one of the first attempts to model the electrophysiological consequences of mechanically activated ion channels in a system known to exhibit mechanosensitivity. Unfortunately, the lack of experimental data prevents rigorous evaluation of the predictions of the model, and therefore the strength of the hypothesis that MAC act as both the sensor and effector system, accounting for the heart's ability to alter its performance in response to stretch.[30]

The ubiquity of MAC in nonexcitable vertebrate cells as well as prokaryotes, fungi, plants, protozoa, and invertebrates suggests a more general function. In addition to mechanotransduction, it has been proposed that stretch-activated channels may be involved in cell motility and regulatory volume decrease in response to hypotonic shock by acting as both sensor and effector (reviewed by Morris).[18]

Inconsistent data

Single-channel studies have indicated the presence of both stretch-activated and stretch-inactivated ion channels in the neuron body and growth cone of *Lymnaea stagnalis* neurons.[31,32] Morris et al[33] reported a failure to elicit neuronal macroscopic currents predicted by single-channel studies by an array of physical stimuli including hyposmotic swelling, stretch, prodding and elevation of intracellular pressure (Fig. 6.5). Exposing the growth cone of the neuron to undefined fluid flow produced a mechanically activated current, but this current was two orders of magnitude smaller than that predicted to result from stimulation of stretch-activated ion channels. These data led researchers to suggest that single-channel mechanosensitivity is an artifact associated with patch recording techniques.

Sadoshima et al reported that cardiac myocytes cultured on collagen-coated silicone membranes demonstrated elevated expression of immediate early genes and hypertrophy as the result of 20% stretching of the membrane (chapter 5).[34] These researchers investigated several potential mechanisms of mechanotransduction, including mechanically activated ion channels. Stretch-activated ion channels were demonstrated in the cardiac myocytes using standard patch clamp techniques (Fig. 6.6, Panel A). The stretch activation of these channels is completely abolished by 10 μM Gd^{3+} (Fig. 6.6, Panel B). Concentrations of Gd^{+3} from 5-50 μM, however, had no effect on the expression of the immediate early genes *c-jun*, *Erg-1*, *c-fos*, or *c-myc*. Furthermore, 10 μM Gd^{3+} had no effect on stretch-induced increases in protein synthesis (Fig. 6.7).

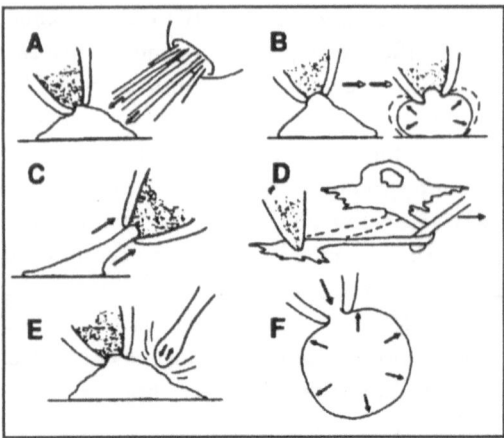

Fig. 6.5. Modes of mechanical stimulation. Shading indicates the perforated patch configuration; solid arrows, force or movement; open arrows, the passage of time. (A) Spritzing of bath solution on growth cone. (B) Hypo-osmotically induced swelling. (C) Stretching of the growth cone by pulling the anchor points at the the substratum and gigaohm seal. (D) Stretching the neurite while recording at the growth cone. (E) Prodding the growth cone with a glass rod. (F) Applying pressure by blowing into the cell. From Morris CE, Horn R. Failure to elicit neuronal macroscopic mechanosensitive currents anticipated by single-channel studies. Science 1991; 1246-9.

These results strongly suggest that Gd^{3+}-dependent stretch-activated ion channels were not the mechanotransducers regulating stretch-induced gene expression and hypertrophy.

The presence of the shear stress-activated ion currents noted by Morris et al in neurons were first observed in endothelial cells. These K^+ currents activated rapidly with flow, did not rapidly desensitize, and quickly returned to basal levels after the cessation of flow.[35] These currents, however, were not observed in outside-out patches exposed to flow.[35] More recently, single-channel recordings have further characterized channel activation due to flow. The channels were not directly exposed to flow (because of the presence of the pipet) but are activated by it,[36] suggesting that they are not directly stimulated by flow. The activation of channels not directly exposed to flow and the lack of response in outside-out channels exposed to flow are consistent with the hypothesis that the putative MAC is opened by a diffusible chemical

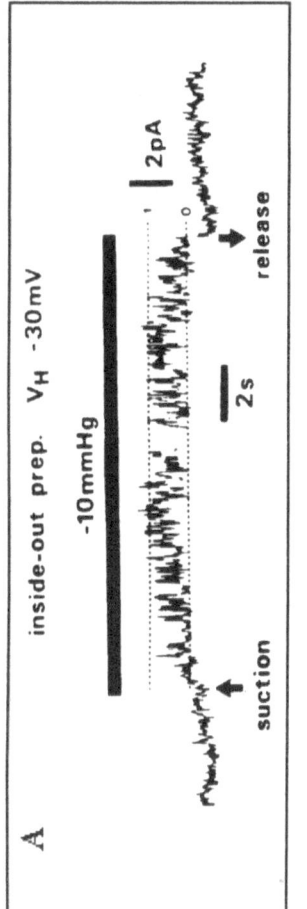

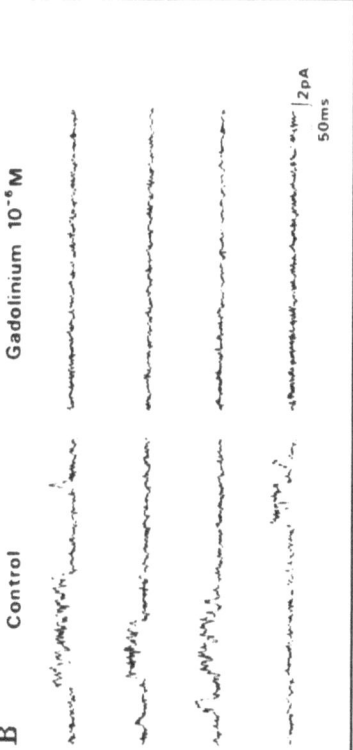

Fig. 6.6. Properties of the stretch-activated channel in neonatal rat cardiocytes. A (top), Typical single-channel recording of a stretch-activated channel current indicates inward current. Baseline artifact is from the mechanical action of suction. In this record, upward deflection of the single-channel current indicates inward current. A broken line indicates closed (o) and open (1) state levels. B (bottom), Effects of Gd³⁺ on the SA single-channel currents. Single-channel currents were recorded in an inside-out preparation, and VH was -30 mV. Minus 10 mm Hg was applied to the pipette, and records are from steady-state suction. Gd³⁺ was applied to the inner side of the membrane; the effect of Gd³⁺ was reversible (data not shown). Upward deflection of the currents indicates inward current. Currents were low-pass filtered at 0.5 kHz. With permission from Sadoshima JI, Toshiyuki T, Jahn L et al. Roles of mechanosensitive ion channels, cytoskeleton, and contractile activity in stretch-induced immediate-early gene expression and hypertrophy of cardiac myocytes. Proc Natl Acad Sci USA 1992; 89:9905-9909.

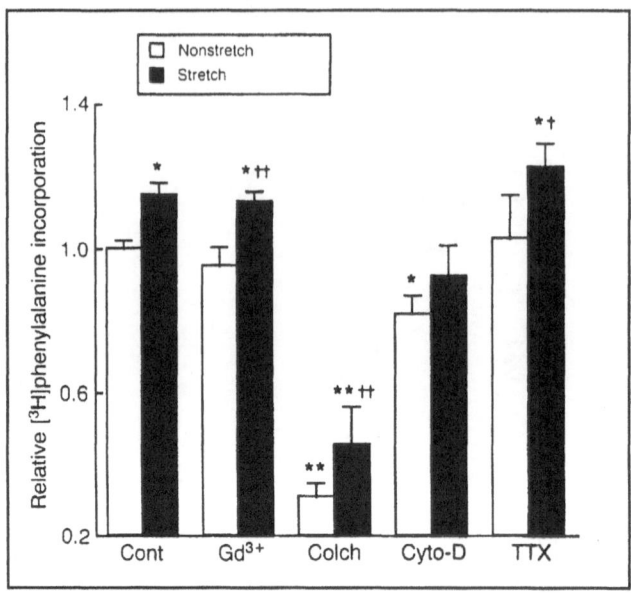

Fig. 6.7. Effect of Gd³⁺, colchicine, cytochalasin D, and tetrodotoxin (TTX) on the stretch-induced increase in ³H-phenylalanine incorporation. Neonatal cardiac myocytes were grown on 20 X 15 mm² collagen-coated silicone sheets. After incubation in serum-free medium for 24 hr, myocytes were stretched (■) in medium containing ³H-phenylalanine (5 mCi/ml) with or without 10 mM Gd³⁺, 3 μM colchicine (Colch), 0.4 μM cytochalasin D (Cyto-D), or 10 mM TTX, for 24 hr. Parallel cultures were incubated without stretch (□) with or without the drugs. Cont, control. Myocytes were harvested with 10% trichloroacetic acid, and trichloracetic acid-precipitable counts were measured. Data are expressed as a counts per dish and are normalized to the mean of control cells without drug or stretch. Data are mean ± SEM from four to seven samples. *, p<0.05; **, p<0.01 vs nonstretch without any drugs; †, p<0.05; ††, p<0.01 vs nonstretch with inhibitor drugs. With permission from Sadoshima JI, Toshiyuki T, Jahn L et al. Roles of mechano-sensitive ion channels, cytoskeleton, and contractile activity in stretch-induced immediate-early gene expression and hypertrophy of cardiac myocytes. Proc Natl Acad Sci USA 1992; 89: 9905-9909.

messenger and is downstream in the signal transduction cascade of the actual mechanotransducer. Two logical candidates for the putative chemical activator of the K^+ channels are Ca^{2+} and NO, because concentrations of both may increase in cultured endothelial cells exposed to flow.[4,7] Endothelial cells possess Ca^{2+}-dependent K^+ channels that, in addition to Ca^{2+}, are activated by NO, at least in smooth muscle cells.[37]

Proposed experiments and commentary

One of the major assets available to researchers of mechanically activated ion channels is the ability to make sensitive (sub-picoamp) real-time measurements at millisecond intervals. Not surprisingly, ion channel activation is often the first measurable event that follows mechanical stimulation; however, just because it is the first measurable event does not prove that nothing precedes it. MAC activation may follow other less easily detected events. For example, the first measurable flow-induced event in cultured endothelial cells is the activation of K^+ currents in patch-clamped whole cells. This current may play an important physiological role, though as discussed above, it likely is secondary to the actual mechanotransducer.

There is little question that certain ion channels, when isolated in a patch clamp, alter their permeability to a range of ions in response to the application of varying pressures within the pipet used for the patch clamp. It is unclear, however, whether forces generated in vivo and in cell culture (other than by the direct application of suction to the plasma membrane) are adequate to affect these channels. Further, it is unknown whether the activation or inactivation of these channels leads to the observed mechanically induced cellular responses (reviewed by Morris[18]). Although there are several reports of inhibitors of mechanically activated ion channels (Gd^{3+},[38] amiloride,[39,40] quinidine,[41] tolbutamide and intracellular ATP,[42] and TTX and diltiazem[38]), a major handicap to researchers is the lack of specific, potent MAC inhibitors. The development of such inhibitors would greatly enhance the understanding of MAC, especially in the study of the role of these channels in physiological systems.

Direct or Indirect Activation of G Proteins

Presentation and consistent data

Perhaps the strongest data implicating G proteins in the mechanotransduction of physical forces are a series of inhibitor studies demonstrating that many, but not all, flow-induced responses in osteoblasts and endothelial cells are inhibited by general (GDP-β-S) and specific (pertussis toxin) G-protein inhibitors. In addition, several flow-induced responses have been shown to be caused by stimulators of G proteins (GTP-γ-S and AlF_4^+). These results are summarized in Table 6.2. In other studies, hyposmotic swelling of myocytes decreases intracellular cAMP concentration, apparently by stimulating G_i.[43]

Care must always be taken when interpreting data from these inhibitor studies. It is possible that basal G-protein activity is required for cellular responses but does not directly mediate responses to physical forces. The fact that many of the flow-induced responses tested can be mimicked by activators of G proteins supports the idea of G-protein involvement in flow-induced responses only if flow can be shown to stimulate G proteins. Direct measurement of G-protein activation is difficult, though new methodologies have been employed. Human umbilical vein endothelial cells (HUVEC) incubated with radiolabeled analogs of GTP were exposed to steady laminar flow or maintained under stationary conditions. After 1 s both sets of cultures were irradiated with ultraviolet light to cross-link the G protein analogs to adjacent compounds. G proteins from cultures exposed to 1 s of flow had a substantial increase in radiolabeling of G proteins compared to time-matched stationary cultures, suggesting that flow rapidly activates G protein turnover (John Frangos, personal communication).

The above studies suggest that fluid flow activates G proteins. It is unclear, however, whether G proteins are the actual mechanotransducers or whether they participate in the signal transduction cascade downstream of the actual mechanotransducers. In the well-studied G-protein coupled receptor systems, G-protein activation follows receptor activation by ligand binding. One hypothesis explaining the involvement of G proteins in flow-induced responses is that flow stimulates the release of autocrine factors that in turn bind to G-protein coupled receptors. One potential (though untested) method to determine the validity of this mechanism is to expose cells to flow in the presence and absence of a cocktail of inhibitors of G-protein coupled receptors. An alternative method to evaluate this hypothesis is to look for G-protein activation in systems lacking functional G-protein coupled receptors. In preliminary studies with G proteins isolated from bovine brain and reconstituted in liposomes (presumably in the absence of receptors), fluid flow stimulated G-protein turnover as measured by the liberation of radiolabeled gamma phosphate from GTP (John Frangos, personal communication). Though these initial results are promising, additional studies are needed to evaluate this hypothesis and corroborate these intriguing results.

Table 6.2. The effects of G protein inhibitors and activators on prostaglandin and nitric oxide production by endothelial cells and osteoblasts.

Cell Type	Product	Condition	No Drug	G protein Inhibitors		G Protein Activators	
				PTX	GDPβS	AlF$_4^-$	GTPγS
HUVEC	PGI$_2$	static	—	—	—		
		acute flow	⇈	—	—		
		chronic flow	↑	—	—		↑
HUVEC	NO	static	—	—	—	⇈	
		acute flow	⇈	⇈	↑		
		chronic flow	↑	↑	↑		
BAEC	NO	static	—	—	—		
		acute flow	↑	—			
Osteoblast	PGE$_2$	static	⇈		—		
		acute flow	↑		—		
		chronic flow			—		

HUVEC, human umbilical vein endothelial cells; BAEC, bovine aorta endothelial cells; PGI$_2$, prostaglandin I$_2$; PGE$_2$, prostaglandin E$_2$; static, no-flow controls; acute flow, laminar flow for up to approximately 1 hour; chronic flow, laminar flow for 1 to 6 hours; —, no change compared to no-flow controls; ↑, increase compared to no-flow controls; ⇈ large increase compared to no-flow controls; PTX, pertussis toxin; GDPβS, guanosine 5'-O-(2-thiodiphosphate); GTPγS guanosine 5'-O-(3-thiodiphosphate).

Inconsistent data

As noted above, Kuchan et al[44] demonstrated that the increase in NO production resulting from exposure to steady laminar flow was inhibited by pertussis toxin. However, chronic (longer than 1 h) exposure to laminar flow resulted in prolonged increases in NO production that were not inhibited by general (GDP-β-S) or specific (pertussis toxin) inhibitors of G proteins. These data suggest that some flow-induced changes in endothelial cells are not the result of G-protein activation, indicating that other mechanotransduction pathways are involved.

Proposed experiments and commentary

G-protein activation in endothelial cells and osteoblasts is necessary for a wide array of responses to laminar flow. It has not been convincingly determined, however, whether G proteins are the elusive mechanotransducers or merely are an essential component in the signal transduction cascade subsequent to mechanotransduction. In light of the similarities between flow-induced and cyclic strain-induced responses in endothelial cells discussed in chapter 2, it would be interesting to determine whether strain-induced responses also are abrogated by G-protein inhibitors.

If G-protein activation follows receptor activation, it will be important to determine which type(s) of receptors are activated and whether this activation is direct or subsequent to other pathways. If G-protein activation is directly caused by physical forces, the mechanism of stimulation or activation must be determined. Another question is whether all G proteins are stimulated by fluid flow or if only specific G proteins are activated. If only specific types of G proteins are activated by physical forces, the identities of these should be ascertained.

Cytoskeleton-Mediated Mechanisms

Presentation and consistent data

As the primary load-bearing structure in the cell, the cytoskeleton is a likely location for mechanotransduction. Wang et al[45] clearly demonstrated that forces applied to the surface of the cell can be transmitted to, and resisted by, the underlying cytoskeleton. These investigators attached magnetic microbeads to β_1 integrins on the surface of a cell, then rotated the beads by applying a magnetic field. The ratio of calculated stress to angular strain (stiffness) increased linearly as a function of stress. This is consistent with the tensegrity model, which defines the mechanical properties of a cell as a system of compressional elements interconnected by elements in tension. The experimentally measured stiffness was decreased by elements that disrupt the cytoskeleton. However, since no physiological or biological response was monitored in this study, it is difficult to determine what role the concepts embodied by the tensegrity model play in the transduction of physical stimuli to biological responses.

Morita et al[46] noted that porcine thoracic aorta endothelial cells exposed to laminar flow with a shear stress of 5 dyn/cm^2 exhibited two temporally related responses: increased endothelin-1 (ET-1) synthesis and increased concentration of globular (G) actin. This rise in G actin was accomplished by depolymerization of filamentous (F)-actin strands; total actin concentration within the cell was unchanged. Based on these observations, they proposed that the increase in G actin was responsible for the rise in ET-1 synthesis. This theory is further supported by the following observations. The F-actin stabilizer phalloidin abolished shear stress-induced increases in ET-1, while actin-disrupting agents cytochalasin B and D mimicked the flow-induced response. Another cytoskeleton-disrupting agent, colchicine, which depolymerizes tubulin, did not affect basal levels of ET-1 mRNA. However, colchicine completely inhibited flow- or cytochalasin-induced ET-1 gene expression. Taken together, these data suggest that the cytoskeleton plays a significant role in the signal transduction mediating endothelial cell response to fluid flow. It is unclear, however, if the cytoskeleton responds directly to mechanical forces or if it is part of the signal transduction cascade.

One system where it appears that the cytoskeleton directly perceives a physical force and transduces this force to a biochemical response is the mitotic spindle of a dividing cell. Whenever a cell divides, it is essential that chromosomes are properly distributed between the daughter cells. As cell division progresses, spindle fibers attach to chromosomes and shorten, pulling the chromosomes to the daughter cells. If a chromosome is not properly attached to a spindle fiber, the process of cell division will be arrested, giving the chromosome additional time to attach. Researchers applying tension to a mitotic spindle with a micromanipulator were able to induce progression into the next phase of cell division (anaphase) even though the chromosome was unattached (Fig. 6.8).[47,48] This experimental evidence suggests that spindle fibers directly sense cytoskeletal tension by an unknown mechanism; if no tension is present, cell division is arrested.

Inconsistent data

Whereas the findings of Morita et al implicate the cytoskeleton as a key component enabling the cell to respond to fluid flow, other data suggest that the degree of polymerization of the cytoskeleton does not control production of NO by human umbilical vein endothelial cells exposed to laminar fluid flow. In these studies, the rate of NO production was inferred by the

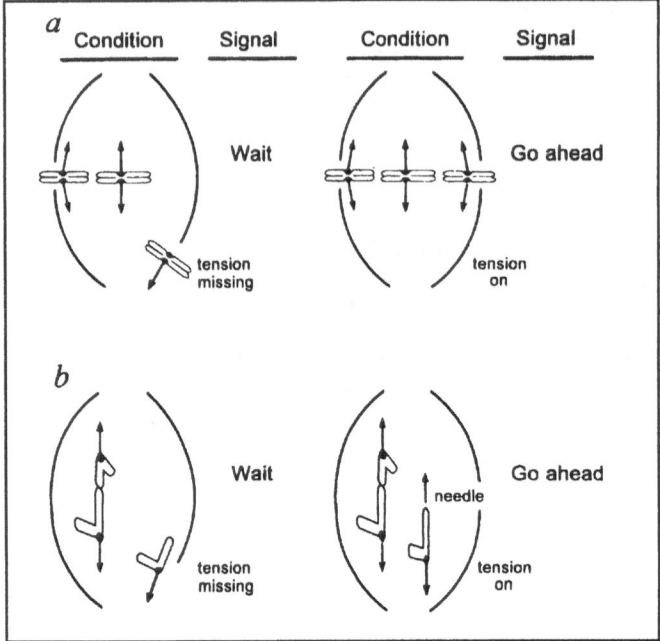

Fig. 6.8. Diagrams of checkpoint control by mechanical force. a, Mechanism proposed for cells in somatic mitosis. Mitotic forces (arrows) act towards a spindle pole from the attachment sites, kinetochores, on each chromosome pair. Properly attached chromosomes are under tension from oppositely directed mitotic forces, but a chromosome attached to only one pole is not. The absence of tension signals the checkpoint to delay cell division. When the last chromosome becomes attached to both poles and is under tension, this signals to the checkpoint that division can proceed. b, Experimental test in mantid cells in meiosis. An X-Y chromosome is not attached: division is delayed. When the missing tension is added with a micromanipulation needle, the cell goes on to division. With permission from: Li X, Nicklas RB. Mitotic forces control a cell-cycle checkpoint. Nature 1995; 373: 630-2.

intracellular concentration of cGMP. Pretreatment of endothelial cells with an actin stabilizer or depolymerizer in concentrations that produced observable changes in the actin filament concentration did not affect the cells' response to shear stress (Heather Knudsen and John Frangos, personal communication). The effects of the drugs on the actin filaments was confirmed by fluorescent microscopy.

As mentioned above, stretched cardiac myocytes increase their expression of *c-fos*. To determine the role of cytoskeletal elements in the mechanotransduction of this phenomenon microtubules were disrupted with 3 μM colchicine and microfilaments were disrupted by 0.4 μM cytochalasin D. The effectiveness of both of these treatments in disrupting the appropriate cytoskeletal elements was confirmed by immunofluorescence. The increased expression of *c-fos* was not influenced by disruption by cytoskeletal elements, suggesting that mechanotransduction does not occur through these elements.[34]

Proposed experiments and commentary

Mechanical forces applied to the surface of the cell can be transferred to the cytoskeleton, which has been shown to be an important component mediating some, but not all, endothelial cell responses to fluid flow. In some systems the cytoskeleton has been implicated as an important component enabling cells to respond to physical forces, but it is unclear whether the cytoskeleton is the actual mechanotransducer or merely is a component in the signal transduction pathway. In one specialized system, the mitotic spindle, the cytoskeleton appears to be the actual transducer.

Summary

One of the most intensely debated questions arising from the study of cellular responses to physical forces is how cells sense these stimuli and convert them into electrical, chemical or biochemical responses, a process known as mechanotransduction. Most types of animal cells studied thus far exhibit responses to physical forces, even those not thought to be subjected to large physical forces in vivo. This suggests that components necessary for mechanotransduction are common to most cells, although the presence of specialized mechanotransducing elements in some cell types cannot be excluded.

Physical forces may be accompanied by modifications of the electrochemical environment, and it is important to determine whether cells respond directly to physical forces or to the electrochemical changes they induce. Some mechanically induced biological effects, such as flow-induced increases in intracellular Ca^{2+}, appear to be initiated by changes in the electrochemical environment adjacent to the cell. However, mathematical models and experimental data suggest that many responses elicited by physical forces result from cells directly sensing those forces rather than associated changes in the electrochemical environment.

Though many hypotheses attempting to explain mechanotransduction have been advanced, none enjoys the support of a substantial majority of researchers, perhaps because no one hypothesis is consistent with the majority of experimental data. This may be because different cell types utilize different mechanisms for mechanotransduction. On the other hand, this inability may reflect deficiencies in the proposed hypotheses, three of which are summarized in Table 6.2.

References
1. French AS. Mechanotransduction. Annu Rev Physiol 1992; 54:135-52.
2. Sachs F. Baroreceptor mechanisms at the cellular level. Fed Proc 1986; 46:12-16.
3. Sachs F. Mechanical transduction by membrane ion channels: a mini review. Mol Cell Biochem 1991; 104:57-60.
4. Dull RO, Davies PF. Flow modulation of agonist (ATP)-response (Ca2+) coupling in vascular endothelial cells. Am J Physiol 1991; 261:H149-54.
5. Mo M, Eskin SG, Schilling WP. Flow-induced changes in Ca2+ signaling of vascular endothelial cells: effect of shear stress and ATP. Am J Physiol 1991; 260:H1698-1707.
6. Kuchan MJ, Frangos JA. Shear stress regulates endothelin-1 release via protein kinase C and cGMP in cultured endothelial cells. Am J Physiol 1993; 264: H150-6.
7. Kuchan MJ, Frangos JA. Role of calcium and calmodulin in flow-induced nitric oxide production in endothelial cells. Am J Physiol 1994; 266:C628-36.
8. Gooch KJ, Frangos JA. Flow- and bradykinin-induced nitric oxide production by endothelial cells is independent of membrane potential. Am J Physiol 1996; 270:C546-51.
9. Busse R. Release of Nitric Oxide. Keynote lecture on endothelial-derived vasoactive factors. Basel, Switzerland, April 1992.
10. Berthiaume F, Frangos JA. Effects of flow on anchorage-dependent mammalian cells: secreted products. In: Frangos JA. Physical Forces and the Mammalian Cell. San Diego: Academic Press, 1993:139-85.
11. Melkumyants AM, Balashov SA, Veselova ES, Khayutin VM. Continuous control of the lumen of feline conduit arteries by flood rate. Cardiovasc Res 1987;21:863-70.
12. Pohl 1991.
13. Hutcheson IR and Griffith TM. Release of endothelium-derived relaxing factor is modulated both by frequency and amplitude of pulsatile flow. Am J Physiol 1991; 261:H257-62.
14. Reich KM, Gay CV, Frangos JA. Fluid shear stress as a mediator of osteoblast cyclic adenosine monophosphate production. J Cell Physiol 1990; 143:100-4.
15. Eriksson C. Streaming potentials and other water-dependent effects in mineralized tissue. Ann NY Acad Sci 1974;289:321-338.
16. Hung C, Allen F, Pollack S, Brighton C. Convective current density in the calcium response of cultured bone cells to fluid flow. 2nd international Conference on Cellular Engineering, 1995, San Diego.

17. Guharay F, Sachs F. Stretch-activated ion channels in tissue-cultured embryonic chick skeletal muscle. J Physiol 1984; 352:685-701.
18. Morris CE. Mechanosensitive ion channels. J. Membr Biol 1990; 113:93-107.
19. Hudspeth AJ. How the ear's works work. Nature 1989; 341:397-404.
20. Pickles JO, Corey DP. Mechanoelectrical transduction by hair cells. TINS 1992; 15(7):254-9.
21. Assad JA, Shepherd GMG, Corey DP. Tip-linked integrity and mechanical transduction in vertebrate hair cells. Neuron 1991; 7:985-94.
22. Hudspeth AJ, Jacobs R. Proc Natl Acad Sci USA 1979; 76:1506-9.
23. Corey DP, Hudspeth AJ. Nature 1979; 281:675-7.
24. Ohmori H. Mechano-electrical transduction currents in isolated vestibular hair cells of the chick. J Physiol 1985; 359:189-217.
25. Sachs F. Baroreception.
26. Yang XC, Sachs F. Block of stretch-activated ion channels in Xenopus oocytes by gadolinium and calcium ions. Science 1989; 243:1068-71.
27. Andersen MC, Yang M. Gadolinium and mechanotransduction of rat aortic baroreceptors. Am J Physiol 1992; 262:H1415-21.
28. Naruse K, Sokabe M. Involvement of stretch-activated ion channels in Ca^{2+} mobilization to mechanical stretch in endothelial cells. Am J Physiol 1993; C1037-44.
29. White E, Le Guennec JJY, Nigretto JM, Gannier F, Argibay JA, Garnier D. The effects of increasing cell length on auxotonic contractions: membrane potential and intracellular calcium transients in single guinea-pig ventricular myocytes. Exp Physiol 1993;78:65-78.
30. Sachs F. Modeling mechanical-electrical transduction in the heart. In: Mow VC, Guilak F, Tran-Son-Tay R, Hochmuth RM. Cell Mechanics and Cellular Engineering. New York: Springer-Verlag, 1994:308-327.
31. Morris CE, Sigurdson WJ. Stretch-inactivated ion channels coexist with stretch-activated ion channels. Science 1989; 243:807-9.
32. Sigurdson WJ, Morris CE. Stretch-activated ion channels in growth cones of snail neurons. J Neurosci 1989; 2801.
33. Morris CE, Horn R. Failure to elicit neuronal macroscopic mechanosensitive currents anticipated by single-channel studies. Science 1991; 251:1246-9.
34. Sadoshima JI, Takahashi T, Jahn L, Izumo S. Roles of mechano-sensitive ion channels, cytoskeleton, and contractile activity in stretch-induced immediate-early gene expression and hypertrophy of cardiac myocytes. Proc Natl Acad Sci USA. 1992;89:9905-9.
35. Oleson SP, Clapham DE, Davies PF. Hemodynamic shear stress activates a K+ current in vascular endothelial cells. Nature 331; 168-70.
36. Jacobs E, Cheliakine C, Gebremedhin D et al. Shear-activated channels in cell-attached patches of aortic endothelial cells. FASEB J.
37. Bolotina V, Najibi S, Palacino JJ, Pagano P, Cohen RA. Nitric oxide directly activates calcium dependent potassium channels in vascular smooth muscle. Nature. 1994; 368:850-3.

38. Ruknudin A, Sachs F, Bustamante JO. Stretch-activated ion channels in tissue-cultured chick heart. Am J Physiol 1993; 265:H960-72.
39. Lane JW, McBride D, Hamill OP. Amiloride blocks the mechanosensitive cation channel in Xenopus oocytes. J Physiol (Lond) 1991; 441:347-366.
40. Hamill OP, Lane JW, McBride DW. Amiloride: a molecular probe for mechanosensitive ion channels. Trends Pharmacol Sci 1992; 89:7462-6.
41. Sigurdson WJ, Morris CE, Brezden BL et al. Stretch activation of a K^+ channel in molluscan heart cells. J Exp Biol 1987; 127:191-209.
42. Van Wagoner DR. Mechanosensitive gating of atrial ATP-sensitive potassium channels. Circ Res 1993; 72:973-83.
43. Hilal-Dandan R, Brunton LL. Transmembrane mechanochemical coupling in cardiac myocytes: novel activation of G_i by hyposmotic swelling. Am J Physiol. 1995; 269:H798-H804.
44. Kuchan MJ, Frangos JA. Role of calcium and calmodulin in flow-induced nitric oxide production in endothelial cells. Am J Physiol. 1994; 266:C628-36.
45. Wang N, Butler JP, Ingber DE. Mechanotransduction across the cell surface and through the cytoskeleton. Science 1993; 260:1124-7.
46. Morita T, Hiroki K, Maemura K, Yoshizumi M, Yazaki Y. Disruption of cytoskeletal structures mediated shear stress-induced endothelin-1 gene expression in cultured porcine aortic endothelial cells. J Clin Invest 1993; 92:1706-12.
47. Li X, Nicklas RB. Mitotic forces control a cell-cycle checkpoint. Nature 1995; 373:630-2.
48. Nicklas RB. How cells get the right chromosomes. Science 1997; 275:632-637.

CHAPTER 7

Production Systems

This chapter will discuss the relationships between mechanical factors and animal cell culture systems used for production systems in biotechnology and tissue engineering. The term biotechnology will be used in this chapter to describe the use of cultured cells to produce macromolecules of potential commercial interest. These macromolecules can range from relatively simple molecules such as penicillin, a molecule of less than 1 kDa produced by cultured fungus, to more complicated products such as insulin (~6 kDa), to extremely complex structures such as the polio viruses (2,500 kDa). Tissue engineering, a newer and less well-defined term, will be used in this chapter to describe applications for which the desired products are either individual cells (such as blood cells for infusion subsequent to chemotherapy) or tissue-like structures formed by cells on a biological or synthetic scaffold (such as bioartificial skin, cartilage, and blood vessel).

Biotechnology

Use of Animal Cell Culture as Biotechnology Production System

Perhaps ironically, the first use of animal cell culture for the production of macromolecules of medical and commercial interest focused on the most complicated of all macromolecules, the viruses. The Salk polio vaccine was licensed and used in human trials in 1954. Over the next 25 years a primary use of animal cell culture was the isolation, assaying, and production of viruses for vaccines. Throughout the 1960s a number of new vaccines were developed, including those for measles, rabies, mumps and rubella. The methodology for producing hybridomas was introduced in 1975, but it was not until 1981 that the first monoclonal antibody diagnostic kit was introduced. In 1982 the first recombinant pharmaceutical, insulin, was synthesized, although microbial culture rather than animal cell culture was used.

It wasn't until 1986, more than 30 years after the first use of the Salk polio vaccine, that animal cell culture, specifically suspensions of lymphoma cells,

Mechanical Forces: Their Effects on Cells and Tissues, by Keith J. Gooch and Christopher J. Tennant. © 1997 Landes Bioscience.

was used to produce the first nonvaccine product, gamma interferon. Recombinant Chinese hamster ovary (CHO) cells were licensed for tissue plasminogen activator (t-PA) production in 1988. Commercial products derived from animal cell culture have been more extensively reviewed by others.[1]

In order to produce this wide array of products, researchers have utilized several different types of bioreactors, most of which fall into one of three classes. The first class of bioreactors closely resembles cell culture systems such as flasks or roller bottles that are used in the laboratory for small-scale production of cells. Scaling up these systems is accomplished primarily by increasing the number of flasks or bottles rather than the size of each individual unit. The second class of bioreactors utilizes anchorage-independent cells such as hybridomas, baby hamster kidney (BHK) or CHO cells grown in suspension. This technique may employ multi-thousand-liter reactors and has been used for the production of very large-scale products such as foot-and-mouth virus vaccine (2×10^9 doses/year). The third class permits large-scale production of desired compounds by anchorage-dependent cells by culturing these cells on microcarriers (normally small beads) which then are suspended.

Origin and Lethal Effects of Mechanical Forces

Mechanical forces that act upon cells result from interactions between the cells and their surroundings. In all systems, the bulk fluid exerts a hydrostatic pressure on the cells. If the fluid is in motion relative to the cell, shear forces and additional pressure forces are generated. In cultures of freely suspended cells or cells grown on microcarriers, fluid motion results in collisions among cells or their carriers with other cells, microcarriers, or immobile surfaces such as the vessel's impellers or walls. The presence of a gas phase introduces additional forces as the result of bubble formation, translation, drainage, and rupture.

The magnitudes of physical forces are relatively small in the first class of cell culture systems. As the volume of the system increases, however, as in the two suspension systems, the magnitude of physical forces also increases. One of the most obvious and widely studied consequences of physical forces in the larger systems is cell death. A theoretical model of cell injury caused by agitation in the absence of entrained or sparged bubbles considers cells and the microcarriers as particles and attempts to correlate cell death with the frequency and/or intensity of particle-to-particle or particle-to-eddy interactions.[2] This model emphasizes the interaction of particles and turbulent eddies of similar sizes, because these interactions result in the transfer of energy from eddy to particle and represent potential cell damage. In contrast, interactions between particles and larger eddies merely result in displacement of the particles without absorption of a significant amount of energy.

A more complicated model that accounts for both particle-to-particle and particle-to-eddy interactions predicts that the rate of cell death resulting from agitation decreases with increasing viscosity. Furthermore, this model predicts a protective effect of increased viscosity that becomes more significant at greater agitation intensities. These predictions of this theoretical model are consistent with experimental data of cell death rates in bioreactors (Fig. 7.1). [3]

In bioreactors where bubbles are present as the result of entrainment of gas from the surface or sparging, they often are the most significant cause of cell death. Photomicroscopy has revealed that animal cells may adsorb onto a bubble as it rises through the bulk fluid, especially as it passes through the liquid-gas interface at the surface of the liquid.[4,5] Theoretical predictions of the forces exerted on a cell adsorbed on a bubble at the moment of rupture indicate that these forces are adequate to cause cell death.[6] These authors

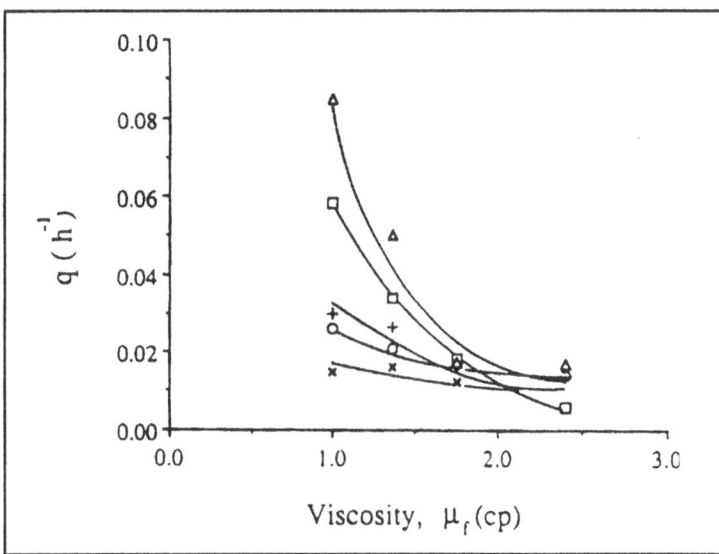

Fig. 7.1. Specific death rate (q) of bovine embryonic kidney cell in spinner flask as a function of viscosity and impeller speed and viscosity. The average difference between specific death rates in duplicate cultures is ~10%. Impeller speeds studied are 100 RPM (X), 120 RPM ( ), 140 RPM (+), 150 RPM ( ), and 160 RPM ( ). With permission from Lakhotia S, Papoutsakis E. Agitation induced cell injury in microcarrier cultures. Protective effect of viscosity in agitation dependent: experiments and modeling. Biotechnol Bioeng 1992; 39:95-107.[3]

developed the following semi-empirical equation for a first-order death rate constant due to bubble translation and rupture in an air lift reactor.

$$k = \Psi \frac{C_f}{C_b} \frac{3hF}{rV}$$

This equation predicts that the death rate, k, is proportional to the volumetric flow rate of gas through the reactor, F, and the thickness of the bubble immediately before it bursts. k is inversely proportional to both the reactor volume, V, and the bubble radius, r. The parameters Ψ (the fraction of cells present in a bubble that are killed when the bubble ruptures) and C_f/C_b (the concentration of cells in the film of the bubble divided by the concentration of cells in the bulk fluid) were determined in independent experiments and were 0.2 and 0.6, respectively, for the system investigated.[6]

Methods of Reducing the Lethal Effects of Mechanical Forces

Cell death due to mechanical forces may be reduced either by decreasing the magnitude, frequency, duration or rate of change of physical forces acting on the cells or by increasing the ability of the cells to resist physical forces. In practice, both of these methods may be achieved by the use of medium additives such as serum, and natural or synthetic polymers. The mechanisms by which medium additives reduce the susceptibility of cells to mechanical forces are not well understood but may include decreasing plasma membrane fluidity,[7] increasing bulk fluid viscosity,[3] or excluding cells from liquid-gas interfaces.[3] The use of medium additives to decrease incidences of death of suspended animal cells due to agitation and aeration has been reviewed by Papoutsakis,[3,8] while Goosen has considered the relevance of this topic in the context of insect cell culture.[9]

One intriguing class of medium additives for reducing the lethal effects of mechanical forces on animal cell culture is perfluorocarbon emulsions. Perfluorocarbon emulsions increase oxygen transfer and medium density, thereby decreasing the level of agitation required to provide adequate oxygenation and prevent cell settling. In addition, these emulsions protect cells from the effects of aeration, possibly by forming a stable layer of foam at the liquid surface that the cells do not enter, thereby separating the cells from bursting bubbles.[10]

Another method of reducing mechanically induced cell death is to modify the operation and design of bioreactors to minimize the mechanical forces generated. Goosen discusses several bioreactors designed to minimize the forces to which the cells are exposed, including air lift bioreactors, hollow fiber bioreactors and vessels agitated with a helical ribbon.[9] The presence of bubbles and the associated forces can be eliminated by the use of membrane aerators, for which a theoretical optimal design has been proposed.[11]

Non-Lethal Effect of Mechanical Forces

The tremendous number of biological responses by cells in culture and in vivo caused by sublethal levels of physical forces noted in the previous chapters suggests it may be inadequate to limit investigation to the lethal effects of forces on cell culture systems used as production systems. Unfortunately, this is almost always done. Knowledge of the effects of mechanical forces on production of biotechnology-related products by cultured animal cells is limited. Reviews exist that emphasize the effects of physical forces on cultured cells before damaging levels are reached, but the limited number of citations of the effect of physical forces on animal cell cultures used in production systems further underscores the lack of research in this field.[12,13]

One of the few studies to investigate the effects of mechanical forces on a cellular response other than death is the work of Lakhotia et al. These researchers modeled the change in the concentration of viable cells in a batch reactor as a function of time with first-order kinetics,

$$\frac{dC}{dt} = Ce^{k_a}$$

$$k_a = k_g - k_d$$

where C is the concentration of viable cells and k_a, k_g and k_d are the apparent growth rate, specific growth rate, and specific death rate, respectively. When CHO cells grown in a 2-liter bioreactor were exposed to excessive agitation, their viability decreased as expected. Interestingly, the specific growth rate, as determined by two-color flow cytometry, also increased with higher levels of agitation. This increase in growth rate persisted for up to 6 hours, even when agitation was returned to basal levels (Fig. 7.2).[14] These data raise the possibility that average apparent growth rate may be increased by stimulating cells by periodically exposing them to elevated agitation followed by a return to basal levels of agitation.

Tissue engineering

Many of the current approaches to tissue engineering follow the same fundamental algorithm. A population of cells is contacted with a synthetic or biological polymer scaffold of the desired shape. The scaffold is seeded with primary or cultured cells and then implanted into an animal. In some cases, the scaffold is implanted in the body almost immediately after seeding, while in other cases the seeded scaffolds are cultured in vitro until desired structural, functional or mechanical characteristics are developed.

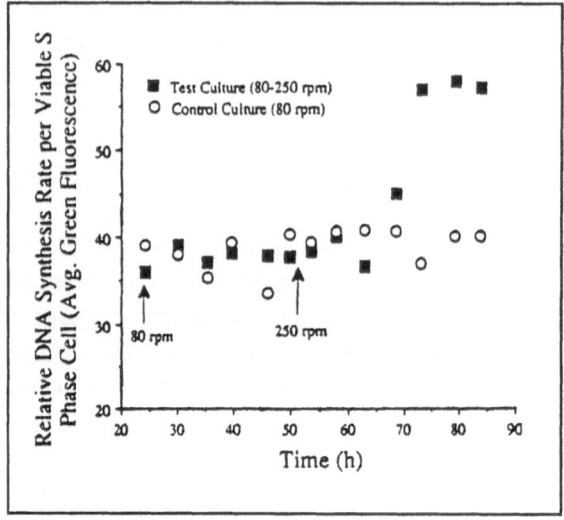

Fig. 7.2. Average DNA synthesis rate per viable S-phase cell
for cultures of CHO cell exposed to control (80 RPM, ○)
and lethal (250 RPM, ■) agitation intensities. Figure 5 of
Lakhotia S, Bauer K, Papoutsakis E. Damaging agitation
intensities increase DNA synthesis rate and alter cell-cycle
phase distribution of CHO cells. Biotechnol Bioeng 1992;
40:978-90.[14]

Alternatively, the cells may be introduced to the polymer scaffold in vivo
by implanting the scaffold in the desired anatomical location and relying on
cell migration from the surrounding tissue for seeding. Both variations of
polymer seeding have been utilized in animal studies and have resulted in
functional tissue.

Tissue Engineering of Bone

State of the art
    Ceramics as well as natural and synthetic polymer scaffolds for osteo-
genic cell transportation in vitro and conduits to promote growth of bone in
vivo have recently been reviewed.[15] Tissue-engineered bone generated by in
vitro seeding of autologous osteogenic cells such as osteoblasts or marrow-
derived stem cells onto suitable scaffolds may reduce the problems of donor
scarcity, immune rejection, and pathogen transfer. A potentially simpler technique to repair bone defects is to implant an appropriate conduit material in
vivo adjacent to bone tissue and rely on migration of cells for seeding.

Role of mechanical forces

As discussed in chapter 3, it has been known for over 100 years that mechanical forces play a prominent role in the development and remodeling of bone in vivo. Bone mass is quickly lost under conditions of diminished mechanical loading such as prolonged bed rest or weightlessness due to microgravity. In light of this widely acknowledged fact, it is ironic that many attempts to develop tissue-engineered bone from cultured osteoblasts and an appropriate scaffold have been conducted under conditions with minimal mechanical forces, conditions that would be expected to result in bone loss in vivo.

One study assessing the effects of mechanical forces on the development and differentiation of osteoblasts in culture produced dramatic results.[33] Rat calvarial osteoblasts were cultured on macroporous collagen beads in Petri dishes (stationary controls) or in a fluidized-bed bioreactor (see Appendix 7.1 for a description of fluidized-bed reactors). The average flow-induced shear stress on the surface of the beads was calculated to be 1 dyn/cm^2 by applying Stoke's Law and ignoring the rotation of the beads. Markers of osteoblast differentiation such as alkaline phosphatase activity and hydroxyapatite (mineral formation) were monitored over a period of 1 month. Analysis of these markers indicated a more rapid and complete differentiation of osteoblasts cultured in the mechanically active environment. Mineral formation as measured by hydroxyapatite levels was dramatically affected; osteoblasts maintained under stationary conditions produced undetectable quantities of hydroxyapatite, while those cultured in a mechanically active environment rapidly began to produce mineral and continued for at least 4 weeks (Fig. 7.3). The implication of these results for researchers attempting to develop tissue-engineered bone is clear; appropriate mechanical forces may be a key determinant of tissue development in vitro.

Tissue Engineering of Cartilage

State of the art

The need for improved treatment for cartilage deterioration, coupled with the potential for generating tissue-engineered cartilage of acceptable quality in the near future, make tissue-engineered cartilage a likely candidate for a successful commercial product. There is a considerable demand for a system that permits the repair of cartilage, as damaged cartilage does not heal well in vivo and normally degrades. Currently there is no effective treatment for mild damage that produces considerable morbidity. Physicians ordinarily will wait until damaged cartilage deteriorates significantly, then, if possible, replace the entire joint with a prosthesis. Prosthetic joints last approximately

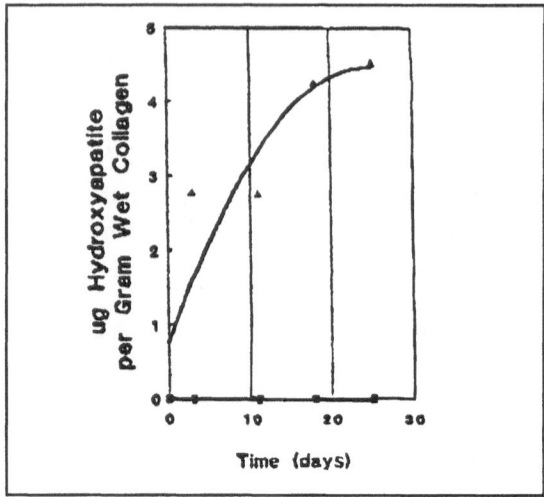

Fig. 7.3 The production of bone mineral (hydroxyapatite) by rat calvarial osteoblasts grown on porous collagen beads is greatly enhanced in the mechanically active environment of a fluidized bed reactor (▲) compared to stationary culture (■). With permission from Hillsleg MV and Frangos JA. Biotechnol Bioeng 1994; 43:573-81.

10 years, at which point they must be replaced. Obviously, such treatment is very expensive and requires the patient to endure considerable pain.

In addition to the clinical demand for improved treatments for cartilage deterioration, several aspects of cartilage structure make it amenable to tissue engineering. Cartilage is nearly avascular and in vivo functions well with minimal nourishment, making nutrient delivery in culture or immediately following implantation less of a concern. In addition, cartilage is a fairly simple tissue composed only one cell type, chondrocytes, and the extracellular matrix produced by these cells.

Several academic and industrial institutions (Genzyme, Cambridge, MA and Advanced Tissue Sciences, La Jolla, CA) are involved in ongoing efforts to develop tissue-engineered cartilage for therapeutic use in humans. These efforts have been reviewed.[16]

Role of mechanical forces

Like bone, cartilage in vivo and ex vivo responds dramatically to applied mechanical loads by increasing extracellular matrix synthesis, which gives cartilage its mechanical properties. Again, like bone, most of the attempts to grow tissue-engineered cartilage have utilized either stationary cultures (Petri dishes)[17] or systems with relatively small mechanical forces such as the National Aeronautics and Space Administration (NASA)-developed rotating bioreactors[18] specifically designed to simulate microgravity or spinner flasks.[17] Cartilage grown in these systems histologically resembles normal cartilage, though its biochemical composition is inferior to native cartilage.[19]

Several studies indicate that the mechanical environment in which chondrocytes are cultured influences the quality of the tissue-engineered cartilage produced. Isolated chondrocytes were seeded on meshes of polyglycolic acid polymers and cultured in Petri dishes or spinner flasks under mixed or unmixed conditions. Tissue-engineered constructs grown under mixed conditions produced cartilage-like tissue with biochemical properties more like native cartilage, with up to 70% more cells, 60% more sulfated glycosaminoglycan (GAG) and 125% more collagen than unmixed controls.[20]

In these mixing studies, however, it is difficult to separate the effects of mixing-induced mechanical forces from those of convection-enhanced transport of nutrients to and metabolites away from the cartilage. In previous studies,[17] the same group demonstrated that increasing the thickness of the polymer scaffold under stationary conditions decreased cellular growth rate. Alternatively, increasing mixing increased the proliferation rate of chondrocytes in scaffold of a given thickness. These two sets of experiments suggest that transport of nutrients or metabolites may limit the proliferation of chondrocytes in tissue-engineered cartilage constructs. If this is the case, it is likely that increased transport due to mixing may account for the observed increases in cell number, which in turn could account for the increased production of sulfated GAG.

To date, all attempts to develop tissue-engineered cartilage have yielded material with inferior biochemical composition and mechanical properties. There are at least three plausible explanations for this. First, there may be a fundamental limitation associated with culturing harvested chondrocytes on synthetic polymers that limits the quality of the final product. Another explanation could be that the maximum time that the tissue-engineered cartilage has been cultured is inadequate to permit proper development; perhaps a longer period of cultivation would produce material more closely resembling native cartilage. A third explanation may be that some aspect of the mechanical or chemical environment in which the tissue is cultured may be inadequate to permit full development of the cartilage. Unpublished work by Gordana Vunjak-Novakovic lends support to the third hypothesis. When pieces of excised bovine articular cartilage were cultured under conditions identical to those used for tissue-engineered cartilage, the concentration of collagen and sulfated GAG decreased with time. Furthermore, the steady state

concentration of sulfated GAG and collagen concentration in the excised cartilage, like tissue-engineered cartilage, depended on the bioreactor used and
mixing conditions employed. These results suggest that some aspect of the
mechanical or chemical environment under which the bioartificial cartilage
is cultured is inadequate to permit ideal development of the cartilage.

If the mechanical environment in which the tissue-engineered cartilage
is cultured is inappropriate, it may be modified to more closely resemble the
mechanical environment in vivo. Systems designed to generate low-amplitude
cyclic loading (such as that experienced by cartilage in vivo) are currently
used to test the effects of mechanical forces on extracellular matrix composition by excised cartilage (see chapter 4 for a discussion of these systems and
the positive effects mechanical loading has on the production of GAG and
collagen synthesis.) These systems may prove ideal for culturing
tissue-engineered cartilage with biochemical composition and mechanical
properties more like native cartilage. Alternatively, it may not be necessary to
develop properties identical to native cartilage, as the material may continue
to remodel once placed in vivo.

Tissue Engineering of Blood Vessels

State of the art

Natural blood vessels have many biochemical and mechanical properties
that an ideal artificial blood vessel must duplicate. Essential properties include a strong and durable yet compliant vessel wall and a nonthrombogenic
luminal surface. An artificial vessel that lacks any of these properties is likely
to experience problems in vivo. Artificial vessels with insufficient initial
strength or those that lose strength rapidly in vivo are prone to aneurysm
and rupture, leading to catastrophic failure. Artificial blood vessels that lack
flexibility promote stenosis of the native blood vessel at the anastomosis, a
phenomenon that is attributed to compliance mismatch between the native
and artificial vessels.

The intimal surface of a successful artificial blood vessel, like the native
endothelium, must be nonthrombogenic. Many vascular prostheses are at
least mildly prothrombogenic. While prostheses for large-diameter arteries
appear to tolerate limited clotting, thrombosis severely limits the application
of vascular prostheses for use in small-diameter (several mm) grafts such as
those required for coronary bypass surgery.

The best replacement for a damaged or occluded artery is an arterial
autograft with a cumulative patency rate of 93% over 5 years.[21] The applicability
of arterial autografts, however, is limited by the availability of donor arteries
of appropriate length and diameter. The availability of donor veins of appropriate dimensions is less limited, and they are frequently used despite their

substantially lower patency (45% over 5 years).[21] The most widely used alternative to autologous vessels are expanded polytetrafluoroethylene (PTFE) grafts, which are limited due to thrombosis. One of the most promising approaches to increasing the antithrombogenicity of synthetic grafts is to seed the grafts with endothelial cells.[22] Studies with 4 mm synthetic polymer grafts seeded with endothelial cells demonstrated a dramatic improvement in patency, from a 14% patency rate over 1 month for control grafts to an 86% rate for endothelial-seeded grafts.[23]

More recently, several different approaches have been taken to develop a bioartificial blood vessel more closely resembling a native vessel in the hope of improving patency. Weinberg and Bell developed a bioartificial blood vessel constructed from collagen and cultured vascular cells.[24] Smooth muscle cells were suspended in a collagen solution that was cast into an annular mold. Over the next several days, the smooth muscle cells contracted the gel around the central mandrel. The contracted gel was removed and its lumina was seeded with endothelial cells while the adventia was seeded with fibroblasts. In this system, the endothelial cells functioned as a permeability layer and produced von Willebrand factor and prostacyclin. The media composed of smooth muscle cells embedded in the collagen gel did not have adequate mechanical strength and needed to be reinforced with Dacron mesh to withstand in vivo levels of pressure (~150 mm Hg).[24]

While the initial publication of Weinberg and Bell reported an attempt to develop in vitro a bioartificial blood vessel closely resembling a native blood vessel with fibroblasts, endothelial cells and smooth muscle cells in the proper location, later attempts provide an appropriate acellular template to promote growth of the appropriate cells in vivo. Organogenesis (Canton, MA) has developed and tested an acellular collagen vascular prosthesis. To minimize blood clotting, the internal surface of the prosthesis is heparinized. When placed in vivo, the surrounding cells grow into and onto the collagen surface. The invading cells remodel the collagen vessel into a structure resembling a native blood vessel. Organogenesis has reported that these vessels remain patent up to 6 months and do not suffer from acute thrombosis even without the addition of circulating antithrombogenic agents.

Role of mechanical forces

The use of synthetic polymer grafts seeded with cultured endothelial cells has been limited by the inability to harvest an adequate number of viable endothelial cells[23,25] and the detachment of the endothelial cells from the prosthesis in the presence of fluid flow with wall shear stresses comparable to those found in vivo. One method of improving cell retention is to precondition seeded grafts by exposing them to flow. Cultured bovine aortic endothelial cells were seeded on 1.5-mm inner diameter spun polythylene vascular grafts and cultured for 6 days in vitro with or without continuous laminar flow.[26] For the first 3 days, flow was such that wall shear stress was ~2 dyn/cm^2, and was raised to in vivo levels of 25 dyn/cm^2 until day 6. Grafts preconditioned with shear and static controls then were exposed to 25 dyn/cm^2 for 25 s. Grafts preconditioned with shear lost ~100 times fewer cells and maintained a confluent endothelium, while the continuity of the endothelium in the grafts not pretreated with flow was severely compromised. Furthermore, the pretreated grafts were less thrombogenic than static grafts, suggesting that culturing cells in the appropriate mechanical environment may substantially increase their in vivo performance.

Under physiological conditions in vivo, arterial smooth muscle cells exist in a nonproliferating, contractile phenotype. Under pathological conditions such as atherosclerosis or vessel injury due to balloon angioplasty, the smooth muscle cells convert from the contractile to the secretory phenotype associated with a loss of contractile elements and an increase in synthetic organelles such as rough endoplasmic reticulum, Golgi complexes and mitochondria. This change in phenotype also is associated with an increase in proliferation.

Smooth muscle cells harvested from native vessels and cultured in vitro readily transform from the contractile to the secretory phenotype.[27,28] Smooth muscle cells undergoing fewer than five cumulative population doublings may spontaneously revert to the contractile phenotype upon growth arrest, a property lost by cells that have undergone more population doublings.[29,27]

The contractile smooth muscle cell phenotype contributes to the strength of the vessel wall and permits it to contract, while the secretory phenotype is associated with hyperplasia and vessel occlusion; clearly the contractile phenotype is desirable, if not required, for any successful bioartificial artery. Therefore, maintaining cultured smooth cells in a developing bioartificial vessel in the contractile phenotype is an important concern.

Recent work by Kanda and Matsuda[30] demonstrates that smooth muscle cells cultured in a type I collagen gel revert to the secretory phenotype, as evidenced by increased synthetic organelles. If these collagen gels are stretched isometrically (static stretch), the smooth muscle cells align parallel to the direction of gel elongation but still transform to the secretory phenotype. If these gels are exposed to periodic stretch (1 Hz, 5% stretch), the embedded smooth muscle cells express the contractile phenotype and align parallel to the direction of stretch.

The data of Kanda and Matsuda suggest a mechanism by which two of the major challenges facing the development of bioartificial blood vessels may be overcome. In vivo, arteries are cyclically stretched by the fluctuating blood pressure associated with the cardiac cycle. If the developing bioartificial blood vessel could be exposed to a similar cyclic strain in vitro, it would be expected that the contractile phenotype may be preserved while causing the smooth muscle cells to align in the correct orientation in the vessel wall. A discussion of the role of mechanical forces in development of a tissue-engineered blood vessel can be found in a review by Ziegler and Nerem.[31]

Experiments using excised vessels demonstrate that vessels cultured in vitro are affected by the environment in which they are cultured. Excised porcine thoracic aortas maintained in organ culture experience smooth muscle cell hyperplasia. After 1 week in culture in the presence of medium 199 with 5% fetal bovine serum, the number of smooth muscle cells in one experiment increased by 42%.[32] No increase in cell number would be expected in vivo, and such rapid proliferation likely would complicate the development of a bioartificial vessel in vitro. Removal of the endothelium at the beginning of the experiment substantially inhibited the increase in the number of smooth muscle cells, reducing it to 15%. Incubation of denuded vessels with medium conditioned by whole vessels restored intimal proliferation (a 30% increase), suggesting that a soluble factor produced by endothelial cells stimulates smooth muscle proliferation. Though the authors did not determine which factors are responsible for this increase in intimal proliferation, a likely candidate is endothelin-1 (ET-1), a potent smooth muscle cell mitogen whose production by endothelial cells is dramatically increased by the removal of physiological levels of fluid flow. If ET-1 or any other endothelial-derived compound whose release is controlled by mechanical forces is responsible for the increase in intimal proliferation, then it may be essential to culture any bioartificial blood vessels under the appropriate mechanical environment.

Summary

Physical forces play a prominent role in regulating the production of compounds by cells in vivo as well as the development and remodeling of tissue in vitro. These phenomena have been extensively studied in animal cell culture to better understand physiological and pathological conditions. Very little research has been conducted, however, to determine the implications of physical forces in cell culture systems used for the production of biotechnology products or engineered tissue. In light of the dramatic effects of physical forces on the biological activity of animal cells in vivo and in vitro, further study of the effects of mechanical forces on in vitro production systems clearly is merited.

Appendix 7.1: Fluidized Beds

Fluidized beds, originally developed and used in the chemical processing industry, also have been used for the in vitro culture of mammalian cells. Anchorage-dependent cells adhere to the surface of the beads and, if space is available, to the pores within the beads. The bed of beads is fluidized (suspended) by the upward fluid flow, producing two well-defined regions within the column: the fluidized bed of beads and the bead-free fluid above the fluidized bed (Fig. 7.4).

Fluidized beads and the cells they contain are exposed to shear stresses and pressures which fluctuate rapidly with time. Additional forces are introduced by the frequent collisions between beads and the walls of the column. Pressure gradients across the beads established by the fluid flow can induce fluid flow through porous beads. This results in the exposure of internal cells to enhanced delivery of nutrients and mechanical forces, though these forces are substantially less intense than those experienced by cells on the surface of the beads.

As discussed in chapter 7, studies with rat calvarial osteoblasts seeded on weighted macroporous collagen beads demonstrated that beads maintained in a fluidized bed exhibit substantially more mineral deposition than beads maintained in stationary conditions. It is not evident from these studies, however, whether the increased mineralization is due to 1) convection induced enhancement of nutrient and metabolite transport within the bead, 2) mechanical stimulation of cells within the bead, or 3) mechanically induced paracrine factors produced by cells on the surface of the bead but acting throughout the bead.

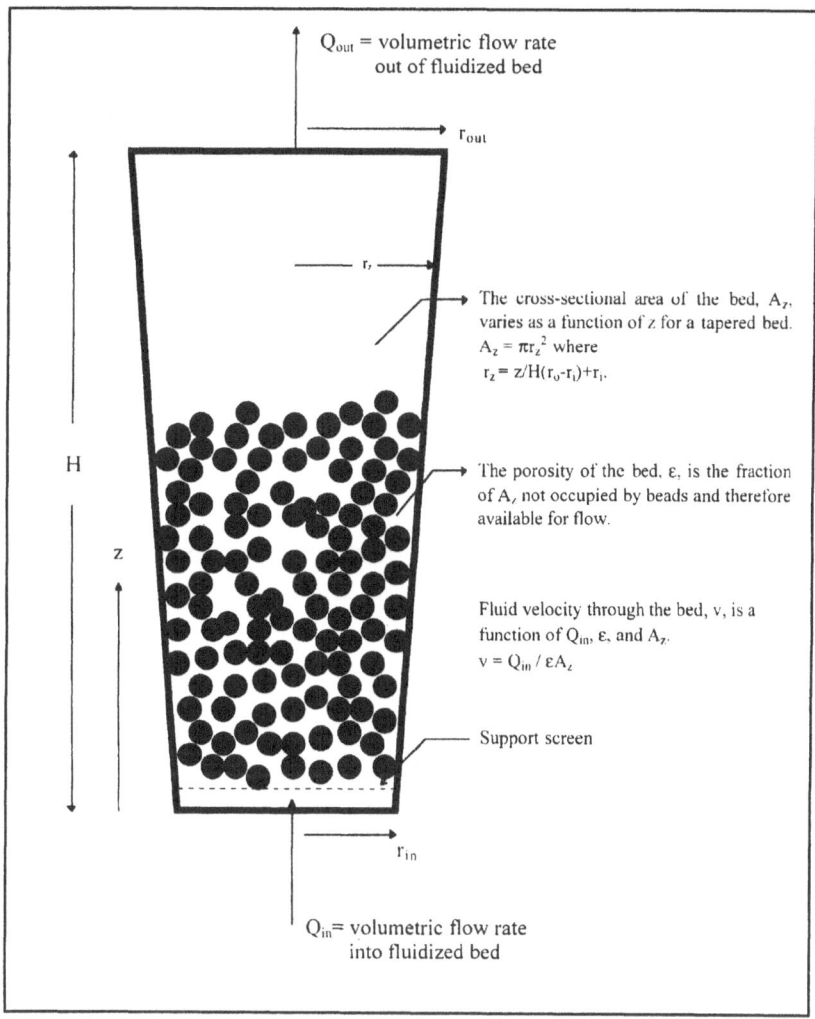

Fig. 7.4. A schematic diagram of a typical fluidized bed system.

Appendix 7.2: Mixed Tanks

As their name implies, mixed tank bioreactors are vessels which are agitated, thereby enhancing transport of nutrients and dissolved gases and preventing settling of cells. Though there are many different variations of this class of bioreactors, ranging in size from less than 100 ml to greater than 100,000 l, they all have the basic structure illustrated in Figure 7.5. Flow patterns may vary, depending on the type and orientation of the mixer, but typical flow patterns are illustrated in Figure 7.5. Researchers have investigated the effects of mechanical forces in these bioreactors on suspended cells[14] and cells attached to microcarriers[3] and macroscopic supports.[20]

Table 7.1. Characterization of the physical and mechanical environment in a mixed tank bioreactor.

Tank diameter	D
Impeller diameter	d
Mixing rate	n
Liquid volume	V
Liquid density	ρ
Liquid viscosity	μ
Impeller Reynolds number	$Re = d^2n\rho/\mu$
Integrated shear factor	$ISF = 2\pi nd/(D-d)$
Impeller tip speed	$T_s = \pi nd$
Dimensionless power no.	$N_p = 0.5$ for $Re > 1000$
Turbulent power dissipation ——————————— unit mass of fluid	$\varepsilon = (N_p n^3 d^5)/V$
Size of smallest turbulent eddies	$\eta = (\mu^3/\rho\varepsilon)^{1/4}$
Velocity of smallest turbulent eddies	$\nu = (\mu\varepsilon/\rho)^{1/4}$

For a more detailed description of these parameters, see article by Vunjak-Novakovic et al[20] and references therein.

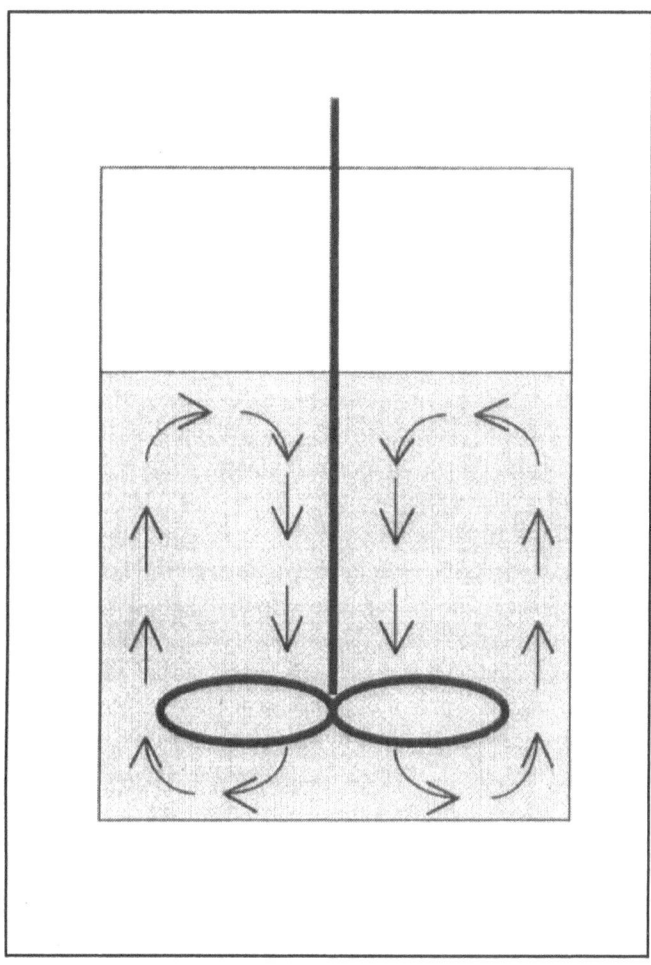

Fig. 7.5. A schematic of a typical mixed tank bioreactor with flow patterns shown.

Appendix 7.3: NASA-Developed Rotating BIoreactors

Developed by NASA to model microgravity, rotating bioreactors consist of an anulus of fluid enclosed in a rotating cylinder (Fig. 7.6, Panels A and B). Some claim that these vessels mimic conditions of actual microgravity; as the fluid element rotates 360° per revolution, the average of the gravity vector is the zero vector after a complete revolution (Fig. 7.6, Panel C). While it is correct that the average of the vectors over a single rotation is the zero vector, the average of the magnitude of the vectors is still g, the magnitude of any single gravity vector. This difference between average vector and average magnitude of a vector can be illustrated by a simple example. Imagine a car driving around a circular track at a constant velocity, v. After each complete lap around the track the car would return to its starting point so that the average of the velocity vectors for one revolution is the zero vector. The magnitude of the velocity, or the car's speed, clearly is not zero. In the same way, the average gravity vector in a rotating bioreactor is the zero vector, but the average magnitude of the gravity vector is the magnitude of g. Under conditions of true microgravity, both the average magnitude of vector and vector are null. In light of the differences between true microgravity and that experienced in rotating vessels, the accuracy of these models is unclear.

In recent years, interest in these bioreactors has increased due in part to a number of NASA grants promoting the use of these bioreactors. As a result of this interest, rotating vessels have found increased application in novel areas such as the culture of macroscopic tissue-engineered structures (see chapter 7). These macroscopic structures either are carried along with the fluid as it orbits the central axis of the bioreactor or, if the drag force on the structure is of the same order as the gravitational force, several alternative outcomes may result. If the gravity force is substantially larger than the drag force, the structure will fall to the lowest point in the vessel. As the drag force increases for a fixed gravity force (e.g., the rate of angular rotation, ω, is increased), the structure will be pushed up the wall of the reactor by the fluid. Further increases in drag will suspend the structure in the fluid.

Researchers have attempted to model the forces on a structure experiencing comparable drag and gravity forces by assuming that the structure is not moving relative to a fixed coordinate system, and therefore the sum of the forces is zero.[18] While this may be a useful first approximation, the authors note that the assumption of no movement by the structure is only an estimate, as tumbling and displacement are clearly evident macroscopically.

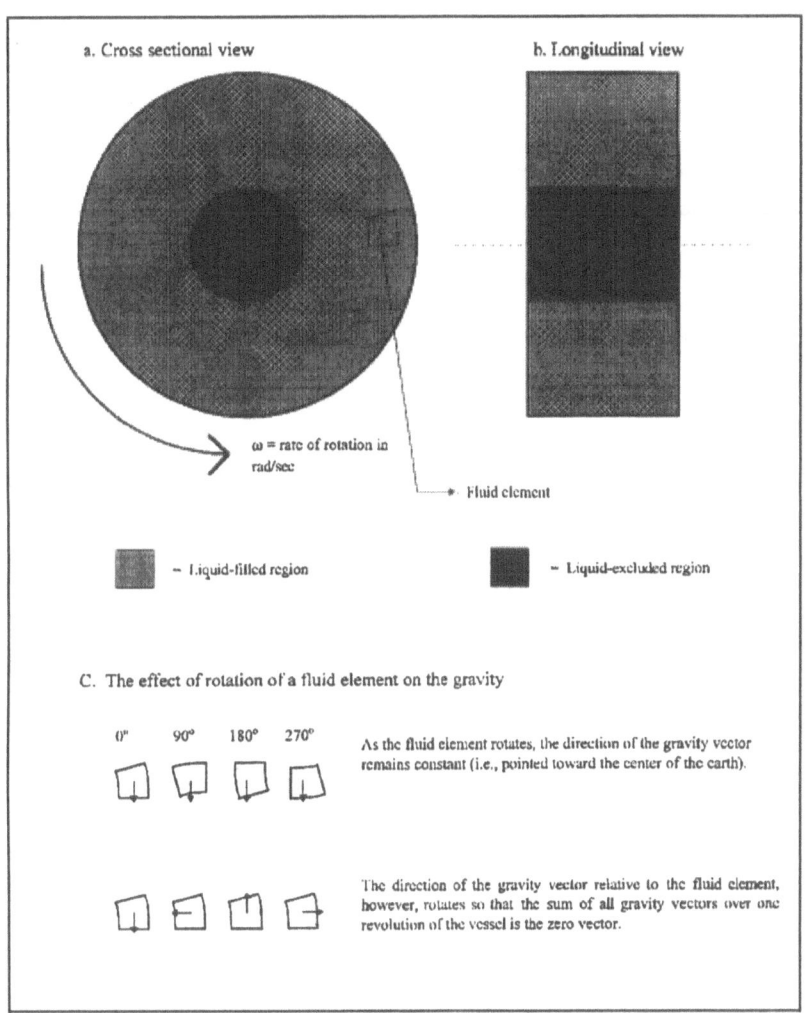

Fig. 7.6. Schematic diagram of a rotating vessel and the effect of rotation of a fluid element on the gravity vector.

References

1. Griffiths B. Products from animal cells. In: Butler M, ed. Mammalián Cell Biotechnology: A Practical Approach. Oxford: IRL Press, 1991: 207-35.
2. Papoutsakis ET. Fluid-mechanical damage of animal cells in bioreactors. Trends Biotechnol 1992; 9:427-37.
3. Lakhotia S, Papoutsakis E. Agitation induced cell injury in microcarrier cultures. Protective effect of viscosity in agitation dependent : experiments and modeling. Biotechnol Bioeng 1992; 39:95-107.
4. Bavarian F, Fan I, Chalmers J. Microscopic visualization of insect cell-bubble interactions. I: rising bubbles, air-medium interface, and the foam layer. Biotechnol Prog 1991; 7:140-50.
5. Chalmers J, Bavarian F. Microscopic visulaization of insect cell-bubble interactions. II: The bubble film and bubble rupture. Biotechnol Prog 1991; 7:151-8.
6. Cherry R, Hulle C. Cell death in the thin films of bursting bubbles. Biotechnol Prog 1992; 8:11-8.
7. Ramirez O, Mutharasan R. Effect of serum on plasma membrane fluidity of hybridomas: an insight into its shear protective mechanisms. Biotechnol Prog 1992; 8:40-50.
8. Papoutsakis ET, Michaels JD. Physical forces in mammalian cell bioreactors. In: Frangos JA, ed. Physical Forces and the Mammalian Cell. San Diego: Academic Press, 1993: 291-346.
9. Goosen M. Large-scale insect culture. Curr Opin Biotechnol 1992; 3:99-104.
10. Ju L, Armiger W. Use of perfluorocarbon emulsions in cell culture. Biotechniques 1992; 12:258-63.
11. Su W, Caram H, Humphrey A. Optimal design of the tubular microporous membrane aerator for shear sensitive cell cultures. Biotechnol Prog 1992; 8:19-24.
12. Merchuk JC. Shear effects on suspended cells. Adv Biochem Eng Biotech 1991; 44:65-95.
13. Gooch KJ, Frangos J. Shear sensitivity in animal cell culture. Curr Opin Biotech 1993; 4:193-6.
14. Lakhotia S, Bauer K, Papoutsakis E. Damaging agitation intensities increase DNA synthesis rate and alter cell-cycle phase distribution of CHO cells. Biotechnol Bioeng 1992; 40:978-90.
15. Crane GM, Ishaug SL, Mikos AG. Bone tissue engineering. Nat Med 1995; 1(12):1322-4.
16. Freed LE, Vunjak-Novakovic G. Tissue engineering of cartilage. In: Bronzind J, ed. The Biomedical Engineering Handbook. Boca Raton: CRC Press, 1995.
17. Freed LE, Vunjak-Novakovic G, Langer R. Cultivation of cell-polymer cartilage implants in bioreactors. J Cell Biochem 1993; 51:257-64.

18. Freed LE, Vujak-Novakovic G. Cultivation of cell polymer tissue constructs in simulated microgravity. Biotech Bioeng 1995; 46:306-313.
19. Freed LE et al. Joint resurfacing using allograft chondrocytes and synthetic biodegradable polymer scaffolds. J Biomed Mat Res 1994; 28:891-9.
20. Vunjak-Novakovic G et al. Effects of mixing on the composition and morphology of tissue engineered cartilage. AICHE J 1996; 42(3):850-60.
21. Lytle BW et al. Long-term (5 to 12 years) serial studies of internal mammary artery and saphenous vein coronary bypass surgery. J Thorac Surg 1985; 89:248-58.
22. Herring M, Gardner A, Glover J. A single-stage technique for seeding vascular grafts with autogenous endothelium. Surgery 1978; 84:498-504.
23. Shindo S, Takagi A, Whittenore AD. Improved patency of collagen-impregnated grafts after autogenous endothelial cell seeding. J Vasc Surg 1987; 6(325-32).
24. Weinberg CB, Bell E. A blood vessel model constructed from collagen and cultured vascular cells. Science 1986. 231:397-40.
25. Herring MB. Endothelial cell seeding. J Vasc Surg 1991; 13:731-2.
26. Otto MJ, Ballermann BJ. Shear stress-conditioned, endothelial cell seeded vascular grafts: Improved cell adherence in response to in vitro shear stress. Surgery 1995; 117(3):334-9.
27. Campbell JHC, Campbell GR. What controls smooth muscle phenotype? Atherosclerosis 1981; 40:347-57.
28. Stadler E, Campbell JH, Campbell GR. Do cultured vascular smooth muscle cells resemble those of the artery wall? If not, why not? J Cardiovasc Pharmacol 1989; 14(6):S1-8.
29. Bjorkerud S. Cultivated human arterial smooth muscle displays heterogeneous pattern of growth and phenotypic variation. Lab Invest 1985; 53:303-10.
30. Kanda K, Matsuda T. Mechanical stress-induced orientation and ultrastructural change of smooth muscle cells cultured in three dimensional collagen lattices. Cell Transplantation 1994; 3(6):481-92.
31. Ziegler T, Nerem R. Tissue engineering a blood vessel: regulation of vascular biology by mechanical stresses. J Cell Biochem 1994; 56(2):204-9.
32. Koo EWK, Gotlieb AI. Endothelial stimulation of intimal cell proliferation in porcine aortic organ culture. Am J Pathol 1989; 134:497-503.
33. Hillsley MV, Frangos JA. Review: Bone tissue engineering: The role of interstitial fluid flow. Biotch Bioeng 1994; 43:575-81.

CHAPTER 8

Overview

The Four Steps of a Mechanically Induced Biological Response

Mechanical forces play an important role in the development, maintenance, and remodeling of tissue in physiological states as well as the initiation and development of disease. Cell culture has been used extensively to study these effects because conditions in vitro are easily controlled. In the preceding chapters, examples of responses to mechanical forces by specific cell types and tissues were considered. In this chapter, we attempt to bring together these individual responses and explore principles common to most or all responses.

To aid in this synthesis, cellular responses to mechanical forces will be broken down into four steps (Fig. 8.1, Panels A-D). These steps were first proposed by Duncan and Turner to describe the response of bone to mechanical loading[1,2] but are applicable to mechanically induced responses in many other biological systems. Please note that Figure 8.1 is not meant to provide full detail regarding the specific response of bone, but rather to illustrate the four steps. In addition, some of the finer details discussed below have not been demonstrated in bone cells specifically, but likely apply to bone as well as to other types of cell.

The first step, mechanocoupling, is the conversion of the applied physical force to secondary forces or physical phenomena detected by the cells (Fig. 8.1, Panel A). Here the application of a force to bone (the primary force) results in the deformation of the bone and interstitial fluid flow. This fluid flow applies a secondary mechanical force, shear stress, to which the cells respond. The second step involves the conversion of either the primary or secondary physical stimulus into an electrical, chemical, or biochemical response (Fig. 8.1, Panel B). This process, mechanotransduction, is an active area of research and debate (chapter 6). The third step, signal transduction (Fig. 8.1, Panel C), is the conversion of one biochemical signal to another. The signal transduction pathways stimulated by physical forces are in many cases identical to those

Mechanical Forces: Their Effects on Cells and Tissues, by Keith J. Gooch and Christopher J. Tennant. © 1997 Landes Bioscience.

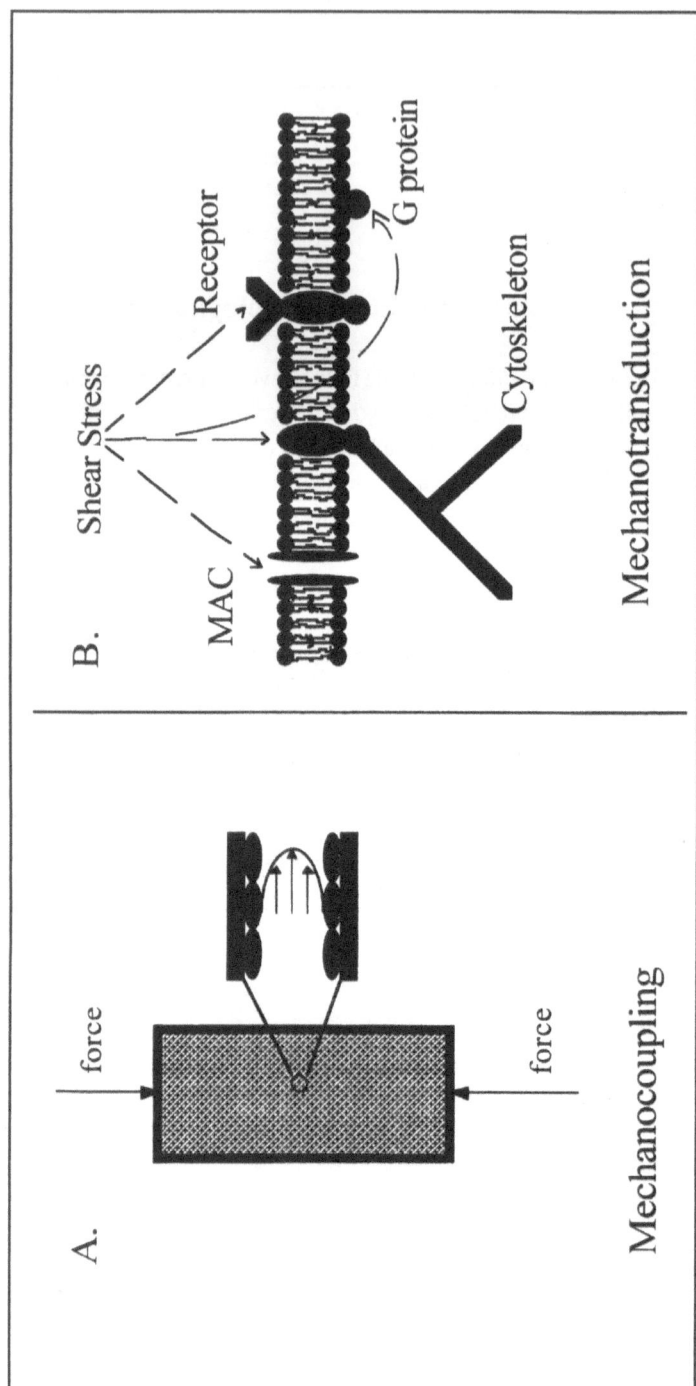

Fig. 8.1A-B. The four steps (continued on opposite page) of a mechanically induced biological response first described by Duncan and Turner.[1] © Keith Gooch, used with permission.

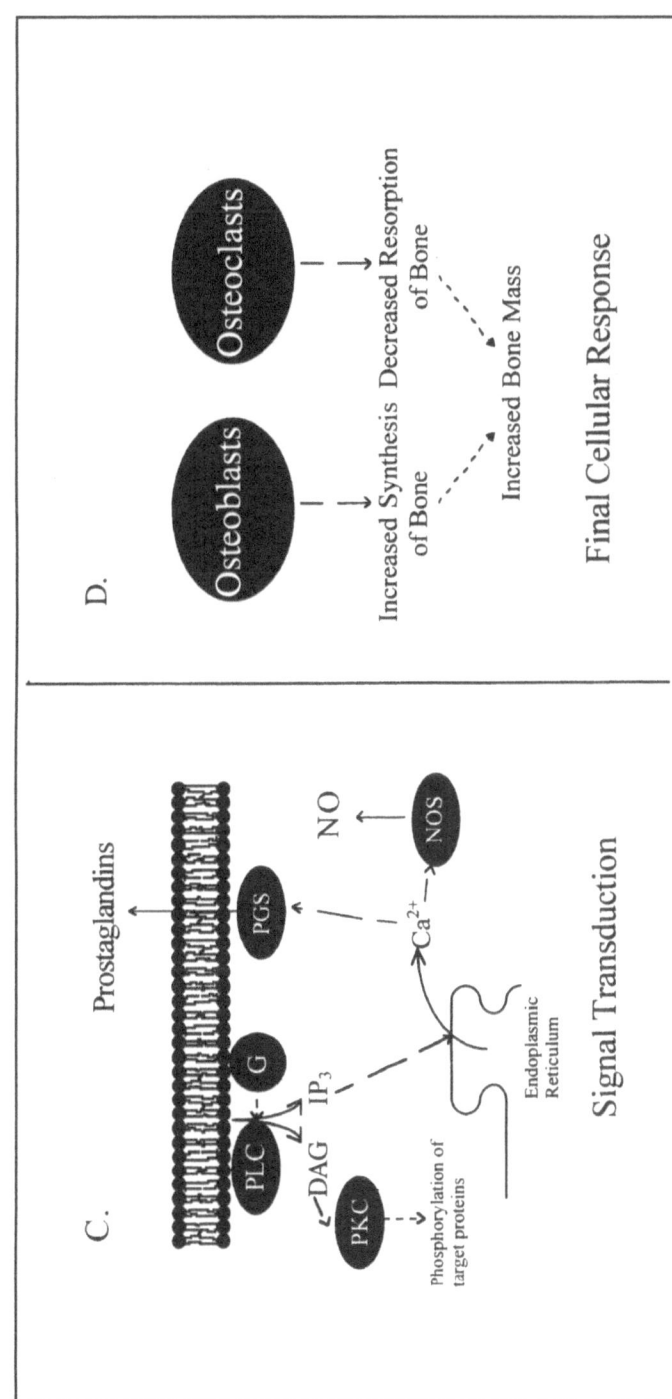

Fig. 8.1C-D.

Table 8.1. The effect of steady and fluctuating flow, strain, and pressure on prostaglandin (PG) production by chondrocytes and endothelial, bone, and muscle cells.

Cell Type	Shear stress		Cyclic Strain
	Steady	Fluctuating	
Endothelial	Human umbilical vein endothelial cells exposed to steady laminar flow with a wall shear stress of ~20 dyn/cm² released a burst of PGI₂ followed by steady release ~8-fold greater than time-matched stationary controls.[3]	Human umbilical vein endothelial cells exposed to steady laminar flow with a wall shear stress of ~20 dyn/cm² released a burst of PGI₂ followed by steady release ~16-fold greater than time-matched stationary controls.[4]	Bovine aortic endothelial cells exposed to cyclic strain for 1, 3, or 5 days and subject to serum free medium supplemented with arachidonic acid for the final 24 hours demonstrated increased PGI₂ synthetic capacity.[5]
Bone cells (osteoblasts, osteoclasts, and osteocytes)	Steady laminar fluid flow with a wall shear stress of 6 and 20 dyn/cm² increased PGE₂ release 9- and 20-fold, respectively.[6]	Chicken calvarial osteocytes exposed to pulsatile fluid flow (5 Hz) increased their release of PGE₂ in a NO-dependent manner.[7]	Bone cells isolated from the calvaria of newborn rats were subjected to biaxial mechanical strains of 0.04% (400 micro-strain) resulting in increased PGE₂ after 5, 15, and 30 minutes.[8]

Chondrocytes	Human and bovine articular chondrocytes were exposed to steady shear stress of 16 dyn/cm². After 48 hours, the release of PGE₂ was increased ~10-fold compared to stationary cultures.
Muscle	Vascular smooth muscle cells exposed to steady laminar fluid flow results in a burst in PGI₂ release followed by continued elevated release. Fluid flow also stimulates PGE₂ production, but only after 3 hours of flow conditions.[9]

Note that the different mechanical stimuli all result in a similar biochemical response in a variety of cell types. While prostaglandin release was chosen to illustrate biochemical response, similar tables could be produced for any of several widely studied responses, such as nitric oxide release or increases in intracellular calcium.

stimulated by biochemical agonists, the primary difference being that the initial biochemical event (in this case, the activation of a G protein) is mechanically induced. The final step (Fig. 8.1, Panel D) completes the conversion from initial stimulus to final cellular response. As a result of the activation of various signal transduction pathways, cellular activities such as proliferation, gene expression, and the release of autocrine and paracrine factors are up- or downregulated, leading to responses on the cellular and tissue levels. Here the cellular responses are increased bone synthesis by osteoblasts and decreased bone resorption of bone by osteoclasts, leading to a tissue-level response of increased bone mass.

The Similarity of Mechanically Induced Biological Responses

The application of a variety of different mechanical forces evokes common or similar biological responses in different cell types. Table 8.1 surveys enhanced prostaglandin production, one of the mechanically induced responses explored in the text. Laminar fluid flow stimulates endothelial, bone, and muscle cells and chondrocytes to enhance their production of at least one type of prostaglandin; in other words, the same force provokes the same response in different cell types. (Note: Flow likely stimulates prostaglandin production by other cell types, but here we will limit our attention to the cell types reviewed in the text). This similarity in response is not true only for mechanically induced prostaglandin synthesis, but for other responses as well. For instance, flow increases nitric oxide production in osteoblasts, chondrocytes, and endothelial and muscle cells. Moreover, in a single group of cell types (bone cells), pressure, strain, steady flow, and pulsatile flow all increase prostaglandin production; in other words, the same effect is achieved by a variety of physical forces. In chapter 2 it was noted that the biological responses initiated by laminar fluid flow and cyclic strain are strikingly similar.

The Coordination of Mechanically Induced Biological Responses

It might seem that if many different mechanical forces stimulate the same response in a number of different cell types, the result of the application of physical forces would be an uncoordinated general stimulation of the cells. However, as was pointed out in the text, tissues respond differently to different mechanical loads. For instance, the acute response of arteries to increased blood flow (i.e., increased shear stress) is vasodilation, while rising blood pressure results in vasoconstriction. While blood vessels, cartilage, bone, and muscle remodel in response to chronic changes in mechanical loading, clearly the responses are very different. These tissue-scale responses result from orchestrated, well-coordinated biochemical responses by cells in direct response to the mechanical forces, cells that are not necessarily the same cells that manifest the final change. For instance, the chronic response to increased blood flow through a vessel is an increase in vessel size, i.e., the vascular smooth muscle cells proliferate and synthesize more extracellular matrix. Though the remodeling occurs via smooth muscle, it is the endothelial cells that apparently sense the changes in the mechanical environment. The intimate linkages between cell types have yet to be fully characterized, but the glimpses we have gotten indicate an astonishing level of cooperation and interdependence.

References

1. Duncan RL, Turner CH. Mechanotransduction and the functional response of bone to mechanical strain. Calcif Tissue Int 1995; 57:344-58.
2. Frangos JA, Eskin SG, McIntire LV et al. Flow effects on prostacyclin production by cultured human endothelial cells. Science 1985; 227: 4693, 1477-9.
3. Sumpio BE, Banes AJ. Prostacyclin synthetic activity in cultured aortic endothelial cells undergoing cyclic mechanical deformation. Surgery 1988; 104:2, 383-9.
4. Reich KM, Frangos JA. Effect of flow on prostaglandin E2 and inositol triphosphate levels in osteoblasts. Am J Physiol 1991; 261:3 Pt 1, C428-32.
5. Klein-Nulend J, Semeins CM, Ajubi NE et al. Pulsating fluid flow increases nitric oxide (NO) synthesis by osteocytes but not periosteal fibroblasts—correlation with prostaglandin upregulation. Biochem Biophys Res Commun 1995; 217:2, 640-8.
6. Brighton CT, Strafford B, Gross SB et al. The proliferative and synthetic response of isolated calvarial bone cells of rats to cyclic biaxial mechanical strain. J Bone Joint Surg Am 1991; 73:3,320-1.
7. Saito S, Ngan P, Rosol T et al. Involvement of PGE synthesis in the effect of intermittent pressure and interleukin-1 beta on bone resorption. J Dent Res 1991; 70:1,27-33.
8. Smith RL, Donlon BS, Gupta MK et al. Effects of fluid-induced shear on articular chondrocyte morphology and metabolism in vitro. J Orthop Res 1995; 13:6, 824-31.
9. Alshihabi SN, Chang YS, Frangos JA et al. Shear stress-induced release of PGE2 and PGI2 by vascular smooth muscle cells. Biochem Biophys Res Commun 1996; 224:3, 808-14.

Index